MW00387893

HORTICULTURAL EXHIBITORS'

HANDBOOK

Cornell University Library

BOUGHT WITH THE INCOME
FROM THE

SAGE ENDOWMENT FUND
THE GIFT OF

Henry W. Sage
1891

A. 184072. 3/1/05

1243

RETURN TO

ALBERT R. MANN LIBRARY

ITHACA, N. Y.

DATE DUE

GAYLORD   PRINTED IN U.S.A.

THE HORTICULTURAL EXHIBITORS'
HANDBOOK.

THE

HORTICULTURAL EXHIBITORS' HANDBOOK

A TREATISE ON CULTIVATING, EXHIBITING, AND
JUDGING PLANTS, FLOWERS, FRUITS,
AND VEGETABLES

BY

WILLIAM WILLIAMSON

GARDENER

REVISED BY MALCOLM DUNN
Gardener to His Grace the Duke of Buccleuch and Queensberry
Dalkeith Park

WILLIAM BLACKWOOD AND SONS
EDINBURGH AND LONDON
MDCCCXCII

T

MANN
SB
98
W73

A.184072

CONTENTS.

PREFACE.

It is well known to all concerned with horticultural exhibitions that a much felt want has long existed of some definite and systematic standard by which the merits of exhibits might be measured with a fair amount of accuracy. In this treatise I have endeavoured to meet this want by giving the details of a simple and efficient system, which can be applied on all occasions, and varied in detail to suit the wants of every case.

A Handbook of concise hints and cultural notes on the rearing and preparation of specimens for exhibition is also a desideratum among horticulturists, especially amateurs and those of limited experience. The cultural remarks in this work will therefore, I hope, be of some practical service to such aspirants to success, and enable them to attain their object with less trouble and more certainty. In the Plant and Cut Flower Divisions of the work I have received valuable assistance from several eminent growers, for which I beg to express my warmest thanks.

Further, I am very specially indebted to Mr. Dunn, Dalkeith Gardens, for his kind assistance in revising the volume; but for all matters of opinion referred to in it, the Author is alone responsible.

In referring to the various genera and species of plants, the names by which they are commonly known in horticulture are generally employed, and no attempt has been made to follow any system of botanical nomenclature.

I trust the work may fulfil the objects aimed at, and form a standard for estimating the merits of exhibits with some degree of precision and regularity, and thus lead to an all-round advance in Horticulture generally.

W. W.

March 1892.

INTRODUCTION.

In introducing the reader to the contents of this Handbook, we consider it necessary to give a few general remarks on exhibitions; to point out some of the benefits they confer on those who participate in them, and show the effect their agency has on increasing the value of produce.

Exhibitions, whether of Science, Art, or Industry, are usually inaugurated with the view of promoting one or more of such desirable objects as the education of the public mind, the gratification of the senses, or the encouragement of Science, Art, or Commerce. That the desired results have been attained in a marked degree in Horticulture is clearly apparent to all observers. When we notice the large and increasing number of shows held throughout the country, nearly every parish and village participating in the annual display of garden produce, and see the interest taken in their success by all classes of the people, we are led to conclude that flower-shows must be doing a vast amount of good, directly and indirectly, in this country. Whatever tends to increase the productiveness of the soil—whether it be the proper application of labour, or the acquirement of a sound knowledge of the wants of plants, and the best modes of applying the elements provided by Nature and Science, to meet the varying demands of plant life—is assisting in a material degree to furnish the means which so largely contribute to the comfort and happiness of the people.

There can be no doubt whatever that exhibitions tend in this direction. The inducements offered through their agency have a stimulating effect on the mental and physical energies of all who participate in competitions, which are in themselves laudable and interesting, when conducted in the right spirit, and with due regard to the rules and regulations imposed by all well-ordered societies. Sometimes these laws are set aside by individuals, who do not hesitate to acquire by illegitimate means any article of which they may be in want; but there can be no real satisfaction to the mind of any one gaining a prize with what is not his own *bonâ fide* production, and the spirit which prompts this dishonourable practice should be combated by all competitors.

It is to be regretted that so few professional gardeners avail themselves of the opportunities for distinction in the art of Horticulture provided for them by the numerous and influential societies throughout the country. Many reasons may be given for this lack of zeal and interest in such competitions. The proprietor may not approve of exhibiting, and in that case the gardener has no choice in the matter. But in a large number of non-competing gardens things might be different if it were generally known how beneficial a reasonable amount of successful competition is to all concerned, especially in the now numerous class of gardens where the surplus produce has to be disposed of in the market. Apart, altogether, from the general effects of the extra attention, application of skill, and the unsparing energy in manual labour, at all hours, which successful competition entails, there is a decided increase in the value of the produce, which is well illustrated by the following incident of actual experience. A few years ago two places were sending equally good grapes to the same fruit merchant, but the money returned to the one was about double that given to the other. An explanation was demanded, and, while admitting that the grapes were equally good, the dealer said the grapes from the one place had got a name in the town, and brought a high price, which was explained by successful competition in the one case and non-competition in the other. A common objection raised to growing fruit for competition is that the plants are under-cropped. This may occur through ignorance or inexperience in reducing the crop to less than what the plant is well able to mature. It should be clearly understood, however, that there is no loss in growing fruit to what may be termed the exhibition standard, that is, the highest state of perfection to which it can be brought. This can only be done by moderate cropping, or just what the plant with good management is able to sustain and perfect. When more than this is attempted, then waste begins, first in inferior produce, and ultimately in the premature sterility and decay of the plant. Another prolific source of non-competition is the belief of many that their place is not adapted for the growth of exhibition specimens; but it should be borne in mind that numerous difficulties have to be surmounted in all cases where growing for competition is carried on, and that if soil and situation be unsuitable for one class of plants, they may be well adapted for the cultivation of others, which may be profitably and successfully grown. Thanks are due to those who come forward with high-class exhibits, and thus keep up the standard of perfection in all divisions of horticultural produce.

Within recent years great excellency has been attained in fruit and vegetable culture. That this has not been altogether the result of improved varieties needs little demonstration. It will as readily be admitted that it is mostly the outcome of an increased knowledge of the best methods of culture, and a better acquaintance with the science which unfolds to us the hidden treasures of Nature.

It is matter for regret that so little has been done till quite recently to promote the study of Science as applied to Horticulture; and while we hear much about the establishment of schools for educating and training youths in the science and art of other branches of rural industry, such as Agriculture and Forestry, very little has been said in reference to the equally strong claims of Horticulture. When the arrangements are being carried out, it is to be hoped that horticultural science will not be overlooked, as it is quite as important a factor in the prosperity of the country and the welfare of the people as any other branch of rural economy. The need for a test-examination in Science and Art is fully as great amongst gardeners as it is among the members of any other branch of the industrial professions. Those who passed successfully would be the better equipped to reap the reward of their industry and natural ability, while others who failed would be relegated to their proper sphere in the horticultural profession. Various associations are doing their best in several parts of the country to make up, in some degree, for this crying want, by offering inducements to young gardeners and others to compose essays on horticultural subjects, for which they are awarded prizes and certificates. These, however, are a very indifferent substitute for a properly equipped curriculum, which would enable them to obtain a thorough knowledge of the theory and practice of horticulture, and thus be the better able to understand and successfully overcome the many difficulties which beset the young and inexperienced.

Horticultural exhibitions are valuable object-lessons to the aspiring but uninitiated tyro, and form an excellent medium for learning what can be attained towards perfection in Horticulture by a diligent application of skill and industry. So far as the circumstances and his knowledge will allow, they furnish to the intelligent student a good criterion of the various points of merit in the different classes of exhibits. This phase of the subject is, however, capable of much improvement. The remark is often heard at flower-shows, "Why has the prize been awarded to this in preference to that?" and such questions may, as a rule, be but the outcome of ignorance on the part of the general public; but the fact must be remembered that the exhibitors are often as ignorant as the others of the relative value of the points upon which the awards have been decided, as no data of any value exist for their enlightenment. One of the chief objects of an exhibition is the all-round improvement of the articles exhibited, which we may safely say has been always kept in view and fairly well accomplished by horticulturists.

Most intelligent people do not now visit flower-shows for the sole purpose of merely admiring gigantic and beautiful specimens, similar in most respects to those they have so often seen on the same occasions in every part of the country, but to scrutinise and note the various points of merit which one exhibit displays over another, and for the possession of which it has been awarded the premier place. In this, however, they as a rule can only find partial enlightenment, as the judges make the awards from certain features in the exhibits which are, we are willing to believe, perfectly clear and intelligible to them, but of which people in general have only the faintest idea, and no means have existed hitherto by which they could be guided in their search for information on the subject.

A simple and efficient method of estimating the value of the merits of all classes of horticultural products met with in competitions, has long been a desideratum to all who are interested in the success of flower-shows, by which an equitable value could be put upon the skill, and taste, displayed in the culture of every exhibit, and the awards made in accordance therewith, on lines clearly defined and intelligible to all who choose to learn. In estimating the value of *Cultural Merit*, which is the principal factor in determining awards to horticultural products, it is really an impossibility to frame a code of laws, or lay down rules which could be followed with success in every case, because no two specimens or samples are found to be exactly alike in every detail. Still by means of the decimal system—a simple method, eminently well adapted for the purpose—a close approach can be made to absolute accuracy in valuing the merits of exhibits, and to practical uniformity in adjudicating thereon.

After summing up the value of the cultural merits displayed by the articles in competition—the result of the thought, skill, and taste of the cultivator—it is not advisable, and is seldom needful, to go farther to find data for determining an award ; but in the event of a tie occurring in the cultural value, resource must then be had to the *Natural Merits* or intrinsic value of the exhibits to determine the award. It is practically impossible to define the relative value under all circumstances of natural merits, because they vary with the tastes and customs of every country and parts of a country, so that what may be highly prized in one place and period may be counted of little use or value at another. Instead, therefore, of attempting to lay down any fixed rules, which in the nature of things could not be of general application, some illustrations are given below of the order of precedence in which the different kinds of exhibits are generally accepted, and which can be easily rearranged to suit the varying taste and circumstances on all occasions. With these desiderata provided, there is still plenty of scope for the exercise of knowledge, acumen, and sound judgment in making a correct adjudication.

It is chiefly in the plant and fruit divisions, where, in mixed collections, the greatest difference in natural merit prevails. The intrinsic, or commercial worth of an article, should not alone determine its relative value in competition, but the greatest weight ought to be attached to the careful perseverance, skill, and taste required to produce a perfect exhibition specimen. For example, the Croton and Palm, which require several years of careful culture to form specimens, should be allowed more competitive value than a Caladium or Coleus; and among flowering plants, the Erica and Ixora should have more weight than a Cineraria or Begonia. A greater difference, however, is generally found in relative value among the fruit in a large collection. The Pine-Apple is usually assigned the first position, closely followed by the Grape, with the Melon, Peach, Nectarine, Fig, Apricot, Pear, Plum, Cherry, Apple, Strawberry, Gooseberry, Raspberry, and Currant in the order in which they are here given. Other less common fruits may intervene, but the order of precedence can always be settled on the spot by competent judges, when they have the exhibits before them. In all cases, excellence for dessert or culinary purposes should carry more weight than mere appearance or rarity. Rich flavour, high quality, and pleasing appearance, should prevail in dessert fruit; in culinary fruits, large size, even outline, fine flavour, and usefulness, are the chief points.

ELECTING JUDGES.

THE greatest care and discrimination should be exercised in selecting the judges for horticultural exhibitions by those whose duty it is to appoint them, because the success of an exhibition depends greatly on the confidence with which the awards are received. A common practice is for a committee, or the managers of a flower-show, to appoint the judges from among their horticultural neighbours, without much consideration as to their qualifications for performing the important duties intrusted to them, in a just and equitable manner. Exhibitors quickly find out any defect in the knowledge, fairness, or want of impartiality in the judges, and promptly resent it, thus leading to much discontent and angry feeling, where nothing of the kind should ever be heard of, if due precautions are taken. Where the selection of the judges is left to the secretary alone, as it sometimes is, the duty may be faultlessly performed by an able and clear-headed man, but, as a rule, this is not a satisfactory method of selecting them. We believe that the greatest confidence would be secured if the judges were nominated by a much wider constituency than is the custom at the present time. It is already the practice at some agricultural and other exhibitions to employ judges who have been nominated for the office by the exhibitors, and this method is generally found to give satisfaction. A similar method could easily be adopted in connection with flower-shows. When the prize schedule is issued to the members, a form should be sent out with it, and all competitors requested to fill in the names of competent judges of the classes in which they intend to compete, and in due course return the form so filled to the secretary of the society. The committee would then only have to find out those who had received the most votes, and elect them

xv   b

to the office. Of course, difficulties may crop up in carrying
out this method ; but where it has been tried, it has generally
resulted in the appointment of judges satisfactory to the main
body of competitors.

The best-qualified judges of the merits of any particular
class of exhibits are, as a rule, to be found among those who
have devoted themselves to the study and cultivation of that
class, and have been successful in bringing it to the highest
state of perfection. It is scarcely possible for any individual
in an ordinary lifetime to acquire a thorough knowledge of all
the merits, peculiarities, and details of every subject brought
under review in horticultural exhibitions; therefore, in judging
the various classes at a flower-show, a division of labour is
absolutely necessary. Some men devote themselves mainly to
florists' flowers, sparing neither time nor means to secure
varieties with the finest form, colour, and substance, and to
cultivate them to the highest possible state of perfection. To
this class of horticulturists, therefore, we naturally look for
the best judges of the merits of the florists' flowers.

Other men are specialists in the cultivation of certain classes
of plants, having by a long course of experience and the exercise of trained skill acquired a thorough knowledge of the
varied wants and peculiarities of the subjects under their
charge, which is needful to ensure success in their treatment,
and the production of perfect specimens at the proper time.
It is obvious that men so trained should be the best qualified
for adjudicating on the various classes of plants.

Then, again, other men find themselves perfectly at home
among fruits and vegetables ; and from long and careful practice in their successful culture and close study of their various
wants, they are thoroughly acquainted with every merit and
peculiarity that fruits or vegetables possess. They have thus
come to be recognised authorities in these branches of horticulture, and should be the best qualified to adjudicate on their
merits.

From the great difficulties in the way of most men acquiring
a perfect knowledge of every department of horticulture, so as
to enable them to adjudicate with confidence and to the general

satisfaction in all, it would probably be more satisfactory for societies to appoint one man as the sole judge in the section to which it is well known he has given special attention, and to strictly confine his duties to that section. One thoroughly qualified man would do the work quite as well, probably better, and more expeditiously than the usual triumvirate now in vogue, and the great responsibility incurred would add to the weight of his decisions.

The very cream of qualified men may be employed to judge, and their awards may be made with the utmost fairness and impartiality, but still their decisions may appear in the eyes of the uninitiated to vary so widely, as to render them a source of much dispute and dissatisfaction. This arises mainly from the absence of any code or recognised system of judging horticultural exhibits. Under the usual arrangements practised by societies at the present day, each arbiter is a law unto himself, and the methods of estimating the merits and demerits of exhibits are as numerous as the judges!

In adjudicating upon the merits of a collection of fruit, for instance, it is a common practice for the judges to begin by allowing *three*, or it may be *four* points as the standard of value of the *best dish* in the collection, no matter how numerous and varied the dishes may be. The highest number of points, say *four*, is given to a first-rate Pine-Apple, *three* to a dish of first-class Grapes, *two* to a fine Melon, and *one* to an excellent dish of Peaches, when the scale is exhausted, and nothing left wherewith to gauge the merits of the other kinds of fruit in the collection. Another common practice is to allow a uniform value to all kinds of fruit, Gooseberries and Grapes being placed on the same level. A much more unsatisfactory method is followed at times, when the judges compare the dishes of the same kind of fruit, and thereby determine the awards by the greatest number of "best" dishes, but taking no account of the aggregate merits of all the dishes in each collection. Worst of all, some judges, after a cursory survey, "slump the lot" in making their awards; and to save their life, they could not give a detailed and satisfactory explanation of the conclusions they have arrived at.

From these familiar illustrations of the crude methods and
unsystematic manner in which judging is practised, especially
at provincial shows, the need for reform will be at once admitted, not only in the *modus operandi*, but more particularly
in the general adoption of a simple and easily applied formula,
by which a fairly accurate and uniform estimate can be made
of the various merits of all classes of horticultural exhibits.
It is in the interest of every Horticultural Society to secure
the services of thoroughly experienced judges.   With local
societies this is often a matter of difficulty, and their best
efforts may fail to command able and experienced men ; but
with a clear and simple method of procedure as a guide, the
merest tyro at adjudicating may, with the exercise of ordinary
intelligence, soon succeed in doing justice to the merits of the
exhibits placed before him for his verdict.

JUDGING.

It should be the aim and desire of all exhibitors to set up their specimens in the most perfect condition attainable, and with all the tasteful effect which their skill and experience can devise, so as to have the best possible chance of winning the highest awards in competition. The merits of every specimen should be brought out to the highest perfection by the skill and taste of the cultivator, and exhibited when they are in their most perfect form. Without going into elaborate details on the specific merits of every class of horticultural exhibits, with all their varied characteristics among plants, flowers, fruits, and vegetables, with which all properly qualified judges should be perfectly familiar, the chief points of merit in each class will be pointed out in its place in this work, when the subject is dealt with.

In ordinary practice, competent judges have seldom any difficulty in arriving at a fair and just conclusion as to the respective merits of the exhibits, after a close examination of the specimens. Where, however, they may differ in opinion, or where the competition is keen and close, the following simple and easy method may be adopted in estimating the merits of the individual specimens, and thereby settling the order of the awards to the satisfaction of all reasonable persons. For example, say nine exhibitors are competing in a class of Twelve "distinct" Stove and Greenhouse Plants, or the same number Rose-Blooms, Dishes of Fruit, or of Vegetables, for which three prizes are offered. On careful examination, it is found that four of the collections are clearly out of the running, and may be passed without further comment. The merits of the other five collections are so nearly equal that it is resolved to estimate the value of the individual specimens to decide the order in which the prizes are to be awarded. Taking them as they come, the collections are marked A., B., C., D., and E., and, taken in the same manner, the specimens in each are numbered 1 to 12; and the value of each article estimated and recorded separately, as shown in the following table. The decimal system is employed as the simplest and best for the purpose, and the values are estimated in tenths and hundredths, the standard being the unit, denoting a perfect specimen.

Nos. of Speci- mens.	A.	B.	C.	D.	E.
1.	·94	·8	1·0	1·0	·9
2.	·8	·9	1·0	·7	·85
3.	·75	1·0	·9	·6	·9
4.	1·0	·8	·75	·77	·8
5.	1·0	·7	·8	·6	1·0
6.	·66	·8	·85	·8	·9
7.	·9	1·0	·7	·9	·95
8.	·7	·82	·99	·85	·7
9.	·6	·7	·8	·8	·8
10.	·7	·9	1·0	1·0	·9
11.	1·0	·58	·81	·8	1·0
12.	·8	1·0	·9	·93	·6
	9·85	10·0	10·5	9·75	10·3

In proceeding with the valuation, A. collection is first dealt with in detail, beginning with No. 1. Each specimen is minutely examined, and the value of its merits agreed upon, and then noted opposite its number in A. column, and so on until the merits of the twelve specimens have been carefully estimated and recorded. The B. collection is treated in exactly the same manner, and the C., D., and E. collections follow in their order. At the conclusion the results are summed up, and show at a glance that C. collection is first, E. collection second, and B. collection third, with the other two clearly in the rear but not far behind, proving that the competition was close, and showing the necessity for minutely estimating the value of every detail. In the rare event of a tie between two or more of the collections, a more minute examination of the specimens in them, and the extension of the decimal to three figures, will to a certainty decide the closest possible com petition.

THE HORTICULTURAL EXHIBITORS' HANDBOOK.

DIVISION I.

PLANTS.

PLANTS constitute the main feature in all horticultural exhibitions, occupying the greatest amount of space, and providing an imposing and attractive display for the admiration of the public. In treating of the various classes of plants, we have taken them in sections with some natural affinities, and in much the same order as they usually appear in the schedules of flower-shows, as follows :—

Section 1. Stove Flowering Plants.
,,   2. Stove Foliage Plants.
,,   3. Greenhouse Flowering Plants.
,,   4. Greenhouse Foliage Plants.
,,   5. Orchids.
,,   6. Ferns.
,,   7. Palms.
,,   8. Table Plants.
,,   9. Miscellaneous Exhibition Plants.

The genera in each section are taken in alphabetical order, for easy reference. Mention is made, when necessary, of a few of the best species and varieties for exhibition purposes. The cultural remarks are confined as much as possible to the special points requiring attention in growing specimens for exhibition.

A

SECTION I.

STOVE FLOWERING PLANTS.

ACHIMENES.

This is a beautiful genus of soft-wooded, tuberous-rooted stove plants, chiefly from the tropical parts of America, which, from their floriferous nature, make excellent specimens, under good treatment, for exhibition. Among the numerous varieties in cultivation, the following are vigorous and free-flowering, and may be chosen for exhibition:—*Carminata splendens, Longiflora alba, L. major, Rosea magnifica, Purpurea elegans,* and *Williamsii.*

The tubercles should be started in a strong heat in a light compost of fine leaf-mould and sand. When the shoots are about an inch long, the plants should be carefully transplanted into pans 12 inches in diameter, placing the plants 2 inches apart, in a mixture of fibry loam and peat, with a third of silver sand and well-dried and broken-down cow-manure, all thoroughly incorporated together. A moist and genial temperature encourages vigorous growth, and a judicious use of weak liquid manure, after the pan is full of roots, induces abundance of fine large flowers. As the shoots grow, they must be carefully staked and trained as near as possible in the shape of a half globe. When the flower-buds begin to appear, the plants should be removed to the warm end of a greenhouse, and be gradually exposed to air and light as they expand.

In preparing them for exhibition, the horizontal side shoots may be untied and carefully drawn towards the centre, and wrapped in cotton-wool and tissue-paper to carry them safe to the show. They must be again neatly arranged and tied-out, taking care to have the plant well balanced, and every flower fully exposed ; all stakes and ties being well hidden by the foliage.

ALLAMANDA.

All the species of this lovely genus of stove climbing plants are of easy culture and remarkably free flowering. They are natives of the tropical regions of South America, and luxuriate in abundance of heat and moisture. Among the best for exhibition are *A. grandiflora*, *A. nobilis*, and *A. Schottii*. To grow them well for this purpose, the shoots should be run up as near the glass as possible till the flowers appear, when they should be brought down and carefully trained over a balloon or other trellis, securely fixed to the pot, taking care to leave sufficient room to allow the leaves and flowers to assume a natural position.

The Allamanda grows well in a compost of equal parts of sound fibrous loam, leaf-mould, and sand, sprinkled with charcoal nodules and half-inch bones. When the specimens are fully formed, and occupy large pots or tubs, it is not necessary to re-pot them annually. With rich top-dressing and a due application of liquid manure, they may be kept in a thriving and floriferous condition for years. Before starting them in spring, the surface soil should be removed to a depth of three inches, and the space filled up with a rich mixture of turfy loam and dry, well-rotted manure, pressed firmly down. The plants require a good rest to ripen their growth, and should be kept comparatively dry through the winter. The shoots should be cut back to a few eyes before the plants are started in spring; and when they break, no more shoots should be allowed to grow than is actually required to furnish the specimen—a few strong and well-matured shoots giving much the finest crop of flowers. Allamandas are usually increased by cuttings which strike and grow easily; but *A. grandiflora* thrives best when grafted on a stronger growing species, such as *A. Schottii*.

If the specimens have to travel a distance to the show, it is very advisable to gum the flowers, by dropping a little good liquid gum on to the base of the petals, which will prevent them being so easily shaken off in transit. Each flower should also be carefully wrapped in tissue-paper and cotton-wool, to save it from being rubbed and injured. When the plants have reached the exhibition, they must be carefully unpacked, and the shoots, leaves, and flowers all neatly arranged in as natural and effective a manner as possible, taking care to leave nothing untidy about them in the way of tying or packing material, and all stakes and ties used in training should be kept well out of view. These remarks as to "packing" and "setting up" of exhibition plants apply to all sections. The finest of specimens are easily ruined by carelessness in packing or in transit to the exhibition; and a prize is often lost by good specimens being set up in a careless, untidy, and tasteless fashion.

AMARYLLIS.

This is a beautiful genus of bulbous plants, the varieties of which have of late been greatly multiplied and improved. There is quite a host of fine varieties in the trade, from which the exhibitor can easily select the most suitable for his purpose.

The practice of some growers is to turn the bulbs out of the pots after they are matured, and dry them off, while others allow them to remain, and start them without re-potting. As the strength of the future plant depends more upon the perfect maturation of the bulb than on the subsequent treatment, the result of the different methods may be imperceptible; but we prefer to allow the bulbs to rest in the pots till the time for starting, when the old soil is shaken away, and three well-ripened bulbs are placed in a 10-inch pot, using a compost of rough fibrous loam, old hot-bed manure, and sand. They should be started in a temperature of 55°, or a little higher, according to the time they are wanted in flower. Little water should be given until growth is proceeding. When the flower-scapes appear, the plants should be taken to the greenhouse, placed near the glass, and watered with weak liquid manure to assist in the full development of the flowers. The flower-stems should be securely staked, the flowers enveloped in wadding, and the leaves carefully laid up against each other, when removing the plants to the exhibition.

ANTHURIUM.

Perhaps *A. Andreanum* and the well-known species *A. Scherzerianum* are the two most popular of this genus with exhibitors, and have a bold and striking appearance in a group of flowering plants. They are of easy culture, and owing to the firm texture of the leaves and flowers make splendid exhibition plants. They luxuriate in a high temperature with plenty of moisture in the atmosphere. During active growth they must have copious supplies of water at the roots, while a little fish-manure dissolved in water and applied once a week during summer will be found a healthy stimulant.

The proper time to re-pot them is early in spring, or when growth has commenced. The pots should be one-third full of crocks, and the compost used should consist of very fibry peat, charcoal, and spagnum ; keep the plants well above the rim of the pots to ensure perfect drainage and a free circulation of air about the roots.

BOUGAINVILLEA.

This is a beautiful genus of stove flowering plants of climbing habit, blooming in summer on the points of the current year's growth. They must be trained near the glass, and have plenty of light to mature their growth. When they show flower, they should be brought down to the trellis, and treated exactly the same as Allamandas. They require to be grown vigorously in summer, and after flowering they should be gradually ripened by being kept moderately dry at the roots, and rested during winter, or till they require to be started to bring them into flower. A little before starting, the shoots should be pruned back to within a few buds of the wood of the previous year. When they break into growth, the plants may be re-potted if necessary, using good turfy loam, leaf-mould, and sharp sand, with comparatively small pots, and potting them very firm. A short time after re-potting the soil should be thoroughly soaked with tepid water, and the plants syringed daily to encourage a vigorous growth. The best for exhibition is the free-flowering *B. glabra*, but *B. speciosa* is also worthy of a place in a collection.

CLERODENDRON.

This genus of plants comprises two groups, the one being scandant in habit, the other shrubby. As *C. Balfourii* is the most suitable species for exhibition, for which it has few rivals, our remarks will chiefly apply to it. Its free-flowering nature, and the ease and certainty with which it can be brought into flower at a given time, makes it invaluable for exhibition purposes. From ten to twelve weeks, according to its season of rest, is the usual period between starting and blooming. The main point to be observed in its culture is to secure strong, well-ripened growth, and then give it a fair amount of rest before starting it again. Being of a climbing habit, the shoots must be regulated during the growing season, and trained up strings or wires close to the glass. When the growth is complete, the shoots should be brought down and neatly trained over a suitable trellis. The greater the number of well-ripened shoots, the more abundant will be the beautiful panicles of bloom. It thrives well in equal parts of loam and peat, with a little old mushroom-bed manure and charcoal. If the loam is heavy, a sprinkling of coarse sand will be beneficial. While the plants are growing freely, weak liquid manure should be given once a week. This plant should not be too often disturbed at the roots, a rich top-dressing being more beneficial when it has attained specimen size.

CRINUM.

Of this handsome tribe of bulbous plants, a good specimen is occasionally seen at an exhibition, and when of a large size and in perfect condition, it is an exceedingly attractive object. Among the best for this purpose is *C. amabile*, *C. Moorei*, and *C. purpurascens*. They thrive best when grown in large pots or tubs, in a compost of fibry loam and peat, with a good sprinking of half-inch bones and charcoal nodules intermixed. The drainage must be perfect; and when the plants are well established, they will not require a shift for several years. A rich top-dressing annually at the time they are started to grow, and a liberal application of clear liquid manure while growth is proceeding and the roots are active, will secure fine heads of the largest-sized flowers. To save the heavy heads of flowers while being conveyed to the show, they must be securely staked, and each flower supported with wads of cotton-wool carefully wrapped with tissue-paper. The plants require to be set in the full sunlight after they have done flowering, and the water gradually withheld, to thoroughly mature the bulbs, as upon their ripeness much of the success in their cultivation depends. They are mostly evergreens, and should not be dried off, like deciduous bulbs, but kept slowly moving through the winter in a temperature of 45° to 50°, and put into stove heat about two months before the date of the show.

DIPLADENIA.

All this genus of lovely stove climbers make excellent subjects for exhibition, and should be grown wherever a suitable house can be devoted to them. They are somewhat difficult to manage, and hence their value as exhibition specimens is fully recognised by all good judges of plants. When their requirements are understood and duly provided for, they can be grown to great perfection.

One important point in their culture is to keep them clean and not over-water them. Although they require a moist atmosphere and high temperature, they will not thrive with a superabundance of water at the roots. Like all other climbing plants for exhibition, they must be trained close to the glass, and should have a corner in the stove where shading can almost be dispensed with, so as to ripen the growth thoroughly. In order to prevent the shoots intertwining, they should be trained singly to strings until they show flower, when they should be brought down and placed over a balloon trellis, on which they should be exhibited. When they cease flowering for the season, water should be gradually withheld, and the plants rested during the winter months. At the same time over-dryness at the root must be guarded against. Three of the most distinct for exhibition are *D. amabilis, D. boliviensis*, and *D. Brearleyana*. The soil best suited for them is a mixture of turfy loam, fibry peat, sharp sand, and charcoal. The pots should be well drained, and the soil pressed firm about the roots.

EUCHARIS.

The lovely and sweet-scented *Eucharis amazonica* is, when well flowered, an invaluable plant for exhibition, as well as of great service for decorative purposes. The genus is a small one, and none of the species are equal to *Amazonica* for exhibition. They are natives of New Grenada, and require a moist stove temperature to bring them to perfection.

To form large specimens for exhibition, select the finest bulbs, wash them clean to see that they are perfectly sound, sprinkle them with silver sand, then plant them rather deeply in pots of a suitable size, placing the bulbs about three inches apart, and using a compost of rich fibrous loam well sprinkled with bone-meal and small bits of charcoal. The pots must be well drained, as the plants require abundance of water at the roots and copious syringing daily to grow them well. The leaves, after potting, should be supported by a few neat stakes till the roots get hold of the soil and are able to support the foliage. A stock of good plants being secured, they can be brought into flower at almost any season by judicious management. To bring them into flower at a given time, they should be encouraged to make a vigorous growth previously, which should be thoroughly matured by keeping them dry and comparatively cool and well aired for about six weeks. They should be taken back to the stove about seven weeks before the show, and treated in the ordinary manner, with plenty of moisture in the air. After the pots are full of roots, the application of clear liquid manure twice a week is very beneficial. Should the flowers come too early for the show, the plants can be removed just as the flowers begin to open to a cooler house, and kept shaded for a week or two till they are wanted. Each flower stem must be securely staked, and the flowers carefully protected in the usual way, for safety in transit to the exhibition.

FRANCISCEA.

This is a beautiful genus of dwarf shrubs from Brazil, which are not so often met with in collections as they should be. They produce an abundance of pretty sweet-scented flowers at different times during the season, and are grown to perfection in a moderate stove temperature. *F. calycina major* and *F. confertiflora* are the best for exhibition. The plants must be kept closely pinched when young, to form a proper foundation for exhibition specimens; afterwards they are easily kept in shape by an annual pruning at the time they have done flowering. They should be re-potted every year in equal parts of fibry peat and loam, with about one-fourth of silver sand, making the whole firm in the pot.

GARDENIA.

This useful genus of free-flowering plants combines the qualities of beauty, fragrance, and profusion of flowers which have made it a popular favourite. Among the best for exhibition are *G. florida intermedia*, *G. f. Fortuneana*, and *G. radicans major*. The short-jointed shoots and shrubby nature of the plant makes it easily grown into a fine compact specimen. It thrives best in a mixture of equal parts sandy loam and peat, with silver sand and leaf-soil; but well-rotted manure and bone-meal should be added to the soil when re-potting specimen plants. It requires a moist and high temperature to make its growth, which should be gradually toned down as the plants come into flower. The Gardenia is very subject to the attack of insects, and should be syringed with soap-water to which has been added a gill of paraffin to the gallon. An excellent preventative for red spiders is to place over the plunging material a layer of stable manure, which should be kept moist. As soon as the buds are set, weak liquid manure with some soot dissolved in it should be given once a week, to increase the size of the flowers, and give a dark, healthy gloss to the leaves.

IXORA.

Although these are considered rather difficult subjects to cultivate, they stand in the front rank of flowering plants for exhibition, for which purpose most of the species and varieties are well suited, especially *I. coccinea superba, I. Colei, I. Duffii,* and *I. javanica florabunda.* They luxuriate in a high temperature, strong bottom-heat, moist atmosphere, and all the light possible, merely shading them from the strong sun at mid-day. While growing, they should be carefully attended to with water at the roots. When done flowering, they should receive any pruning they require, and be rested, in a comparatively dry state, in an intermediate house. They succeed best in fibry peat, a little leaf-mould, and plenty of silver sand, all well mixed, and potted moderately firm. After they have reached specimen size, they will thrive well for years without re-potting; but they should be carefully supplied with weak liquid manure during the season of active growth.

JASMINUM.

There are many species of this excellent genus of trailing sweet-scented plants. Of the stove kinds, *J. gracillimum* and *J. Sambac flore pleno* are the best. The former is a profuse bloomer, and the shoots being very flexible, it can be grown as a specimen in any form to suit the taste of the cultivator. The balloon or the pyramid shape is best adapted for an exhibition specimen. They thrive well in a mixture of sandy loam and peat, with the addition of some old rotted manure. When the plants attain a large size, their luxuriance is very much promoted by applications of weak liquid manure during the growing period.

MEDINILLA.

Among the finest and most handsome of all stove plants is a well-grown and profusely-flowered specimen of *M. magnifica,* a native of the Philippine Islands. It delights to make its growth in a high temperature and a moist atmosphere, shaded from strong sunshine. When the growth is fully developed the plant should be exposed to all the light possible, to thoroughly ripen the wood, and then rest it in an intermediate house. From eight to ten weeks before the plant is required for exhibition, place it again in the stove. An open compost of turfy loam and peat, with a sprinkling of nodules of charcoal and half-inch bones, and plenty of silver sand to keep all open, grows it well. When the roots have filled the pot, and especially when the flowers are developing, a weekly application of clear liquid manure has a very beneficial effect. Cuttings strike readily in pure sand, with a strong bottom-heat, and covered with a bell-glass. The large and beautiful rosy-pink panicles of flowers remain in perfection for a considerable time ; but, in removing the plant to the exhibition, each panicle must be carefully supported by stakes and cotton-wool, as their weight is apt to damage the flowers if they are allowed to sway about or rub against anything.

PANCRATIUM.

This is a genus of bulbous, sweet-scented, white-flowering plants of some merit for exhibition. *P. fragrans* is a free-flowering plant, with large umbels of flowers of perfect purity, and long narrow segments radiating from the centre, which gives it a chaste and graceful appearance. It should be potted in strong fibrous loam, sand, and well-rotted hot-bed manure. It is generally cultivated under the same conditions as the Eucharis with great success. The plants, if placed in an intermediate house previous to flowering, with plenty of light and air to mature the growth, will continue much longer in flower than when confined to the stove. A few doses of liquid manure after the flower-scapes appear are productive of good results, but they should be stopped when the first flowers begin to open. When removing the plants to the show, the greatest care is necessary to prevent injury to the flowers, which should be enclosed in wadding without pressing against it, and the whole plant enclosed in tissue-paper and covered with stout canvas.

POINSETTIA.

This is one of the most striking and attractive winter decorative plants we possess. The typical variety *P. pulcherrima* and the white variety *P. pulcherrima alba* are both very showy plants, but the former is the most suitable for exhibition. A two-year-old plant, if the wood is well matured, makes a good foundation for an exhibition specimen. The plants should be pruned to within two eyes of the old wood, and although they grow fairly well in a cold frame during the summer months, they are all the better of a little artificial warmth, especially during the night. When the buds are started, the old soil should be shaken away from the roots, and the plants potted amongst sandy loam and leaf-soil. They should be syringed twice a day, and very little water given at the roots until growth has fairly commenced. They should be kept near the glass, and plenty of light and air given, for the formation of strong, firm, short-jointed wood. From ten to twelve strong shoots are sufficient to form a fine specimen. Flowering may be forwarded or retarded by the time the plants are introduced to stove heat, three to four weeks only being necessary for the full development of the bracts after the plants are full grown.

RONDELETIA.

The Rondeletias are hard-wooded evergreen shrubs from the West Indies, producing freely their bright little trusses of flowers. Among the best are *R. cordata* and *R. speciosa major*. The roots are small and fibry, and are liable to be injured by injudicious watering; but when the plants are grown into well-flowered specimens, they are very effective at the autumn shows when stove flowering plants are scarce. On account of their straggling habit they require to be neatly trained into shape for exhibition; and in whatever form they are trained, the shoots should be fully exposed during the growing season, that they may be short-jointed and well matured. After flowering, the shoots should be cut back to within three joints of the previous year's growth. When they are starting to grow is the best time to re-pot, using a mixture of rough fibrous peat, light loam, and sharp sand, with a few bits of charcoal. Firm potting and careful watering are two essential conditions to ensure success.

STEPHANOTIS.

The fine stove-climber *S. floribunda* is indispensable to the exhibitor. Its pure white waxy flowers and thick leathery leaves fit it admirably for withstanding the changes of temperature which plants in exhibitions are subjected to.

Under good cultivation it is a rampant grower, and should be trained in the same manner as other climbers, by running up strings under the glass around which the shoots will twine. They can then be easily brought down to the trellis when they show flowers. Although a plant of rapid growth and a gross feeder, it flowers best when rather confined at the roots; therefore large plants can be grown in comparatively small pots, and they will flower more freely if judiciously treated with liquid manure. The best time to re-pot is when they are starting into growth in spring, using good fibrous loam, leaf-mould, sand, and a little bone-meal. Careful watering is very essential, especially during the winter months, when the plant is resting. During the period of active growth, an occasional application of weak liquid manure will greatly benefit established plants which are getting pot-bound. The balloon form is the best mode of training for the Stephanotis, which should be done some time before the exhibition, so that the leaves and young shoots may have time to assume a natural position.

TABERNÆMONTANA.

This is another sweet-scented useful stove evergreen shrub, somewhat resembling the Gardenia. *T. coronaria flore pleno* and *T. grandiflora* are among the best for exhibition, the flowers of the former being pure white, and the latter of a yellow colour. They form by careful pruning and pinching fine compact bushes, which, when well flowered, stand high among exhibition plants. They thrive well under the same conditions of culture as recommended for the Gardenia.

THUNBERGIA.

This is a genus of stove and greenhouse climbers, some of which only are worthy of cultivation as exhibition specimens. *T. laurifolia* is a large-flowered species of a blue colour, which, from the length of time it continues to bloom, entitles it to be classed amongst exhibition plants. It makes a fine-looking plant when it is trained on a globular trellis with all its panicles of flower turned outwards. It requires a compost of fibrous loam and peat in equal parts, with a little sand and well-decayed manure.

TOXICOPHLÆA.

Among stove-flowering plants *T. spectabilis* occupies a useful position, its white scented flowers being produced with great freedom. It is of a rather straggling habit, but with good treatment and carefully pinching and regulating the growth, it can be made to form a fine specimen for exhibition. It grows well under ordinary stove treatment, and should not be over-potted. When the pot is full of roots, weak liquid manure should be applied once or twice a week in the growing and flowering period.

VINCA.

The stove species of this genus, *V. rosea*, and *V. r. alba* are excellent free-flowering plants, which, when well grown and full of flowers, are very attractive specimens for exhibition. The stems and branches being very flexible, they can be trained into any form. An intermediate temperature suits them best, and they should be potted in light sandy loam with a sixth of well-decayed manure.

JUDGING.

In judging stove flowering plants, the chief points of merit are: (1) Profusion, quality, and freshness of the flowers; (2) large size, brilliancy of colour, and sweet odour when present; and (3) size, health, vigour, and cleanness of the plant.

SECTION II.

STOVE FOLIAGE PLANTS.

ACALYPHA.

A beautiful and easily grown genus of foliage plants, which form fine specimens for exhibition when carefully pinched and regulated in their growth. Among the best for this purpose are *A. macrophylla* and *A. musaica*. A compost of peat and loam in equal parts and a small portion of silver sand grows them well. They thrive best in a moist stove, where they should be kept near the glass, to obtain plenty of light to fully bring out the beautiful colours and markings of the leaves.

ALOCASIA.

This is a highly attractive family of plants, with bold, massive foliage, often curiously shaped, and beautifully marked and coloured. They make grand exhibition specimens when well grown, and are very effective plants in any collection. Some of the best are *A. macrorhiza variegata*, *A. metallica*, *A. Thibautiana*, and *A. Veitchii*. To grow them with the greatest success, they require a high temperature, moist atmosphere, and abundance of water at the root while growing. The pots should be about half-filled with drainage, and the compost be of rich, light turfy loam, peat, and chopped sphagnum, freely sprinkled with nodules of charcoal and sandstone. Clear liquid manure is very beneficial when the roots have fully occupied the soil.

ANTHURIUM.

The large, handsome, and distinctly marked leaves of some of the Anthuriums are very conspicuous and telling objects in a collection of foliage plants, especially when grown and exhibited without a blemish. *A. magnificum*, *A. Veitchii*, and

A. Warocqueanum are among the best and most distinct for
exhibition. They require the same culture and treatment as
the Alocasias; giving them liberal supplies of clear liquid
manure when they are growing freely, and keeping the air
thoroughly saturated with moisture, so as to induce development in the foliage to its fullest extent.

ARALIA.

The Aralias are a very distinct and ornamental class of
foliage plants, and the finer forms, such as *A. Chabrieri*,
A. elegantissima, A. Reginæ, A. Veitchii, and *A. Veitchii gracillima* are in great favour with exhibitors of plants for table
decoration. They should be grown in light loam and peat,
adding a little leaf-mould and sand, according to the texture
of the loam. Vigorous young plants produce the finest foliage,
and a succession of these should be kept up by cuttings struck
in the usual manner, or by grafting the weaker on to the
stronger growing kinds. The scion must be neatly and securely
fitted to the stock, and they should be kept in a close moist
atmosphere until union is complete. Perhaps the best method
of securing well-furnished small plants is to notch the stem
of a leading shoot at the base of the healthy and perfect
foliage, tying a handful of moss over the notch. Keep constantly moist, and roots will soon push out, and when the moss
is full of them, take off the shoot with the moss and roots
intact, and pot them in a light, sharp compost. Keep closely
shaded till the roots are through the soil, and then gradually
expose to light and air. By this method well-furnished
specimens are quickly formed, with perfect foliage from the
surface of the pots.

CALADIUM.

This is an easily grown and useful class of exhibition plants,
the luxuriant growth of their brightly and diversely coloured
foliage, forming an excellent contrast with the greenery of
other subjects. It is easy to make a selection for exhibition
from among the numerous fine varieties now in cultivation,

some of the oldest and best known making grand specimens in a short time under generous treatment. They thrive best in rough fibry loam and peat and silver sand, sprinkled with nodules of charcoal and bone-meal. Abundance of water, moist air, and shading from bright sunshine assist in the production of the finest leaves ; and when the pots become full of roots, liquid manure should be freely applied. When the leaves attain their full size, a drier and more airy atmosphere deepens the colours and makes the plant more hardy for exhibition. After the leaves ripen and die off in the autumn, the pots should be laid on their sides in an intermediate temperature ; and special care must be taken to keep them at rest without risk of becoming "dust dry," as the roots are then very liable to become "mealy" and worthless. They are easily propagated by division of the roots, and specimens of any size are readily formed by placing a sufficient number of strong roots or tubers in a pot of the desired size.

CARLUDOVICA.

An elegant and useful genus of foliage plants, with a resemblance to palms in their habits and appearance. *C. elegans* and *C. palmata* make the best specimens. They grow freely in a moist stove, in a mixture of two parts peat and one of loam, with a liberal sprinkling of silver sand. A copious supply of liquid manure, when the plants are growing freely, increases the size of the leaves and the elegance of the plant for exhibition.

CISSUS.

In *C. discolor* we have one of the most beautiful of all the variegated-leaved plants in cultivation. It is a free-growing stove climber, and makes a grand specimen for exhibition when brought out in its best condition and neatly trained over a suitable trellis. It thrives well in a compost of rough turfy peat and loam, with a little old mushroom-bed manure and sand, all well incorporated. When the pot is filled with roots, an application once a week of clear liquid

B

manure improves the size and brightness of the colour of the foliage. Being a deciduous plant, it should be kept rather dry during the winter, receiving just sufficient water to keep the wood from shrivelling. It should be pruned, re-potted, and started in a moist stove about three months previous to the show, so as to have the foliage fresh and perfect. Vigorous young specimens produce the finest leaves; and as the plant is readily increased by cuttings, a succession of plants is easily maintained.

CROTON.

Almost the whole of the numerous varieties of variegated-leaved Crotons make first-rate exhibition plants, and most of them can be grown into fine specimens in a comparatively short time. They take a high place among foliage plants, and when well grown and their rich colours brought out to perfection, they are objects of great attraction. They grow luxuriantly in a high temperature, with a close, moist atmosphere; and to obtain that, the ventilators require to be kept rather close, admitting air by the side ones only, and shading the roof to prevent the sun scorching the plants. In such a temperature the plants make the largest leaves and freest growth. When the plants have made their growth, the shading should be gradually dispensed with, and all the light and air possible admitted, while keeping up the high temperature, to colour the foliage to perfection. The plants should be thoroughly drenched with the syringe in the afternoon in summer, and in the forenoon in winter. A good syringing with soap-suds—two ounces of soft soap dissolved in a gallon of water is sufficiently strong—once a week is beneficial to keep off insects, as well as to give the leaves a clean, healthy appearance. They succeed best in a turfy loam of a medium texture, with a little bone-meal, nodules of charcoal and sandstone added according as the nature of the loam requires them. The pots must be well drained, and the plants must never be allowed to get dry at the root, which immediately affects the foliage and spoils it. Large specimens growing in pots full of their roots may have an occasional supply of weak liquid manure, but care must be taken not to give it too often or too rich, lest the foliage begin to lose colour. Notching and mossing the stem, as described for Aralias, is the best method for obtaining well-furnished small plants.

CYANOPHYLLUM.

When grown to perfection, *C. magnificum* forms one of the grandest specimens for exhibition among all foliage plants. It thrives best in a high moist temperature, in a compost of turfy loam and peat, freely sprinkled with nodules of charcoal, sandstone, and half-inch bones. It must have perfect drainage, as it delights in abundance of water while growing, but it must not be allowed to stagnate about the roots. When the plant is growing freely, and the pot is well filled with roots, copious supplies of clear liquid manure should be given twice a week, which will grow the leaves to their largest dimensions. On strong well-grown plants they should reach a length of about 4 feet, and 2 feet wide. Special attention must be paid to this, and all other large-leaved fine foliage plants, to see that insects do not effect a lodgment on the under side of the leaves, where they quickly multiply and do much harm. Frequent application of the syringe, and the careful use of weak soap-suds in syringing, help much to keep the foliage free from insects, clean, and healthy. The Cyanophyllum should always be grown to a single stem, as the large and finely-marked leaves are displayed with the best effect on such plants. It strikes freely from cuttings of side-shoots taken off with a heel, inserted in sandy peat, and placed in a strong bottom-heat under a bell-glass. Young plants, when well managed, produce the finest leaves, and in a season will make splendid specimens for exhibition.

DICHORIZANDRA.

A pretty genus of stove plants, the flowers of which are often as attractive as the handsome and beautifully marked foliage. Among the best for exhibition as foliage plants are *D. metallica picta*, *D. mosaica*, and *D. vittata*. They thrive best in well-drained pots, in a mixture of equal portions of fibry loam, peat, and leaf-mould, with a sprinkling of silver sand. A moist stove temperature and copious supplies of water at the root, with daily syringings and shade from bright sunshine, grows them well. After the pots are well filled with roots, clear liquid manure may be given once a week; but it must not be overdone, to cause rank growth and loss of colour in the foliage. They require to have a thorough rest during winter, but they must be kept in the stove and water withheld, as they do not bear cold with impunity.

DIEFFENBACHIA.

A free-growing and very effective genus of fine foliage plants, which, under liberal treatment, produce fine exhibition specimens in a comparatively short time. Some of the finest for this purpose are *D. Baraquiniana*, *D. Bausei*, *D. Jenmanii*, and *D. Weirii superba*. Being natives of the tropical parts of South America, they require a high temperature and abundance of moisture to grow them well. They delight in rich open compost, of turfy loam and peat, with leaf-mould and sand freely intermixed, and a little dry cow-dung rubbed down amongst it. After the roots have reached the side of the pots, they are much benefited by frequent doses of clear liquid manure.

DRACÆNA.

Like the Croton, the Dracæna is indispensable to the exhibitor of foliage plants. Numerous varieties, with richly coloured leaves and graceful habit, have appeared during recent years, among which it is easy to select good kinds for exhibition. They thrive well in rich fibry loam, with a little peat, sand, and nodules of charcoal, and should be grown in an atmosphere saturated with moisture and shaded from bright sun, to fully develop the leaves. With abundance of moisture in the air, no syringing should be necessary, and as a rule they are finer without it. Dracænas are extremely easy to propagate, as every inch of the stem will root and grow, inserted in sand in a strong bottom-heat. Tall plants, with bare stems and good heads, should be notched and mossed immediately below the leaves; and as soon as the moss is full of roots, take off the tops and pot them with the moss intact, shading them in a warm moist place till they are well established, and they will soon make nice useful plants.

FICUS.

This genus provides many useful plants which are more or less ornamental, the most noteworthy being *F. Parcelli*. As an exhibition plant it is one of the most striking and attractive. It has also a good habit, and can with very little trouble be grown into a symmetrical specimen. It requires an open compost of loam and peat, and an abundance of water while growing. It should be syringed often to prevent the attacks of red spider, which is its worst enemy. It should be placed near the glass, so that the ivory-white blotches may be well brought out on the leaves. The strongest shoots must be pinched, and the others tied into position where necessary, to form a symmetrical well-furnished pyramid. As it approaches the resting period, water should be gradually withheld, and the plants placed in an intermediate temperature.

HIBISCUS.

The only one of this genus suitable for cultivation as a fine foliage plant for exhibition is *H. Cooperi*. It forms a fine specimen, and in variety of colour it equals, if it does not excel, any other foliage plant in cultivation. In its young state it should be potted amongst fibry loam, peat, sand, and leaf-mould, but when it has attained to a large size, soil of the poorest description should be used in potting—starving it, as it were, into colour. It requires a strong heat and a position close to the glass to bring out the various rich colours in their greatest perfection.

MARANTA.

A genus of useful and easily cultivated plants; but being natives of moist warm regions, they require careful handling to withstand the dry cool atmosphere of exhibition tents. Among the best kinds are *M. Lindenii* and *M. Veitchii.* To grow good specimens and bring out the full beauty of their leaves, a high temperature and moist atmosphere are absolutely necessary. They thrive well in a mixture of fibry peat and loam, with a small proportion of silver sand and a sprinkling of bone-meal, and should receive copious supplies of water at the roots when growing. Keep them rather dry in the stove to rest them during winter.

MUSA.

A genus of large-leaved noble-looking plants, most of which grow far too tall for ordinary stoves; but *M. Cavendishii*, *M. coccinea*, and *M. zebrina* are well adapted for culture as exhibition plants. Their cultivation is simple and easy if they are kept clean, and a proper degree of heat and moisture maintained. Under such conditions they soon rush up into large specimens, their handsome leaves producing quite a tropical appearance wherever they are placed. They should be potted in rough rich loam and sand, and when pot-bound they should be heavily top-dressed with rich manure.

PANDANUS.

A handsome genus of very graceful foliage plants, with leaves mostly of a green colour; but *P. javanicus variegatus* and *P. Veitchii* combine elegance with a fine light variegation, and may be considered the best for exhibition. Young shoots of the purest and slenderest character, springing from the base of the stem, should be taken off with a heel, potted in sandy soil, and plunged in a propagating-house, where they will soon take root. They should be shifted on as they require it, but using small pots in comparison to the size of the plants. When they have reached specimen size, they should be potted in coarse sandy loam, with plenty of pieces of sandstone and charcoal. They should get all the light possible without injury from bright sunshine, to produce perfect variegation. When removing the plants to the exhibition, the leaves should be tied up to the centre, and carefully wrapped round with cloth, to prevent their spines doing injury to other plants placed near them during transit.

PAULLINIA.

The best of this genus for exhibition is *P. thalictrifolia*, the habit of which is scandant, but by close pinching it can be made to assume the bush form, when it very much resembles a specimen of some of the exotic ferns. It forms the best exhibition specimen, however, when it is trained on a balloon or pyramidal trellis. It grows freely when potted in light loamy soil and leaf-mould, kept in a moist warm atmosphere, syringed daily while growing, and supplied with weak liquid manure when the pot is filled with roots.

PAVETTA.

There are few species of decorative foliage plants that equal *P. borbonica* in attractiveness, when its beautifully marked leaves are exhibited in their best condition. It is closely allied to Ixora, and succeeds well under similar treatment. The soil best suited for its growth is fibry peat and loam, with a few half-inch bones, and about one-fourth of silver sand. The plant should be repeatedly pinched while young, to make it branch out into a bushy specimen, which it does slowly. After growth is complete, it should be well exposed to the sun, to bring out the beauty of the markings on its leaves.

PHILODENDRON.

A large genus of tropical plants, of easy cultivation, and among which are some noble-looking specimens for the purposes of the exhibitor. Few foliage plants can compare with *P. Andreanum* for stately effect, and *P. Lindenii* and several others are nearly as striking and effective. Natives of the warmer parts of America, they luxuriate in a high moist atmosphere, with a rich open soil to grow in, and copious supplies of water both at the roots and overhead. When the pots are full of roots, liquid manure may be freely given with the best effect. They are easily propagated by cuttings of the strong stems, which quickly root and form fine plants.

RHOPALA.

An elegant genus of fine foliage plants, some of which are very useful for decorative purposes, as well as furnishing beautiful specimen plants for the exhibitor. Among the finest are *R. corcovadensis*, *R. elegantissima*, and *R. Vervœneana*, which all grow best in a stove temperature, although they also thrive well in a warm greenhouse. A compost of fibry loam and peat, with charcoal nodules and sand, grows them well. They delight in moisture while making growth, but should be gradually exposed to air and light, and, placed in an intermediate temperature, they will retain their foliage in a perfect condition through the whole year, and are thus nearly always available. They strike freely from cuttings placed under a bell-glass in a strong bottom-heat, or the tops may be notched and mossed, when good-sized plants are secured at once.

SANCHEZIA.

The only species of this genus suitable for exhibition as a foliage plant is *S. nobilis variegata*. This is a widely cultivated plant, which grows freely, and makes a handsome specimen when its leaves are fully developed and their variegation well brought out. It should be potted in equal parts of fibry loam and peat, with some small bits of charcoal and silver sand. It must be grown near the glass and shaded only to prevent scorching, and have the flowers nipped out as they appear. The leaves should be frequently sponged to keep them clear of aphides, which prove harmful to them when in a young state.

SONERILA.

This is a genus of beautiful dwarf-growing plants, as much admired for the beauty of their flowers as for their foliage. The best for their foliage are *S. margaritacea Hendersonii*, *S. m. marmorata*, and *S. m. superba*, which all possess a branching, compact habit, and form very neat specimens. They bloom freely, and the flowers look well; but, like all plants when grown specially for their foliage, the flower-buds should be nipped out as they appear. A compost of two parts fibrous peat and one of chopped sphagnum, sand, and charcoal nodules grows them well. They should receive plenty of water and shade in summer, but be well exposed to the sun and kept slightly dry when resting.

SPHÆROGYNE.

The plants belonging to this genus are highly ornamental, and among the best for exhibition are *S. imperialis* and *S. latifolia*. Under proper treatment their leaves grow to a large size, and form noble-looking plants; but to secure perfect foliage the plants must be grown in a moist warm temperature and well shaded. They very much resemble the Cyanophyllum, and thrive well under the same treatment.

TERMINALIA.

The only species of this genus suitable as a foliage plant for exhibition is *T. elegans*. Its persistent habit of growing with a single stem makes it rather difficult to work into a large specimen. It thrives well in a mixture of loam and peat with about a fourth of silver sand. To produce a large specimen, the plant should be cut down while young, and repeated if necessary to secure four shoots, which should be bent outwards and staked, to keep them clear of each other. A native of Madagascar, it luxuriates in a warm moist atmosphere, with plenty of water when growing, but should be kept moderately dry in winter. It is much benefited by a judicious application of liquid manure, like many other foliage plants; but care must be exercised in its use, as leaves with rich delicate colours are liable to be spoilt if the plants receive too much feeding.

THEOPHRASTA.

A genus of stately foliaged plants, with large coriaceous leaves of great substance and beauty. Of those in cultivation,

T. imperialis and *T. macrophylla* form grand specimens for exhibition. Natives of Brazil, they luxuriate in a strong moist heat while making their growth, with a copious supply of water at the root and overhead. When growth is well ripened, they stand in an intermediate temperature a long time in fine condition. They thrive well in a mixture of fibry loam and peat, with a good sprinkling of half-inch bones and sand; and while the plant is growing and the roots active, frequent doses of clear liquid manure encourage the development of the largest-sized leaves.

TILLANDSIA.

An extensive genus of beautiful stove plants, most of which are epiphytal, although under proper treatment they thrive well in pots and baskets. Among those cultivated as foliage plants, most of which also produce lovely flowers, the best known are *T. musaica*, *T. splendens*, and *T. zebrina*. They are an easily grown class; but while they are making growth they thrive best in a high moist temperature, with a copious supply of water at the root and heavy syringing overhead daily. The compost they delight in is a fibry loam and peat, with a sprinkling of half-inch bones, and nodules of charcoal and sandstone, in well-drained pots. They should be kept rather dry while at rest during the dormant season.

JUDGING.

In this class above most others, large size is of small value, unless accompanied by all-round excellence. Under-sized foliage plants are also of little value, however well done they may be. A useful medium size, according to the nature of the plant, with every part fully developed, is always preferable.

The properties of foliage plants are: (1.) Condition, habit, symmetry, and size of the specimen; (2.) the foliage healthy, clean, and perfect; and (3.) the colour and variegation of the leaves should be clear, vivid, and in good contrast.

SECTION III.

GREENHOUSE FLOWERING PLANTS.

ACACIA.

A numerous genus of free-flowering and highly ornamental plants, chiefly natives of the temperate regions of Australia, and thriving well in a cool greenhouse. *A. armata*, *A. Drummondi*, and *A. Riceana* are among the finest species for exhibition. They grow well in equal parts of light turfy loam and peat, with plenty of silver sand and a sprinkling of sandstone and charcoal nodules. The plants should be cut back after flowering, and with the aid of a few degrees more heat, and a thorough syringing overhead twice daily, they will soon break into growth, when they should be re-potted. Give as small shifts as possible, and make the soil quite firm in the pot, so as to induce sturdy and more floriferous growth. The drainage must be ample, so as to allow copious supplies of water to be given during growth, and when the roots have fully occupied the soil a few doses of liquid manure are very beneficial. They are apt to make rampant growth under liberal treatment, and the strongest shoots should be regularly pinched to form a good bushy specimen. Such straggling kinds as *A. Riceana* may have the shoots run up the roof of the greenhouse on strings till the flowers appear, when they can be neatly trained over a suitable trellis for exhibition. Acacias are very hardy greenhouse plants, and after they have made their growth and set their flower-buds, they may be placed out of doors, along with other greenhouse plants, to ripen their wood, and they will flower the better for it next season. They must be taken back to the greenhouse before any danger from frost arises.

ACROPHYLLUM.

A large specimen of *A. venosum*, when full of its beautiful flower-spikes, has a charming effect among other plants in a collection. Its pinkish-white flowers are so very distinct from all others that it is always a welcome subject at summer shows. It should be potted in the beginning of March amongst fibrous peat and about one-fourth of sharp sand, and kept in a cool airy place in the greenhouse. Being of a compact habit, it requires very little tying or trimming, but it must be kept clean by syringing daily during the summer. It should be well supplied with water at all times, and a little liquid manure, given two or three times before flowering, tends to develop the flower-spikes to their fullest extent and beauty.

AGAPANTHUS.

The blue African lily is a plant that suits itself to almost any treatment, and being of a hardy nature and tenacious of life, its culture is often neglected. It is not a first-class exhibition plant, but its graceful arching leaves, independent of its fine flowers, make it a good associate amongst stiffer-growing plants. Two of the best varieties are *A. umbellatus flore pleno* and *A. umbellatus maximus.* They should be potted amongst rough loam, sand, and decomposed manure, and receive an abundance of both clear water and liquid manure in summer, but should be kept drier in the winter season.

APHELEXIS.

An old-fashioned genus of greenhouse plants, of a rather straggling habit of growth, but which by skilful treatment make very beautiful and attractive exhibition specimens. Among the finest for that purpose are *A. macrantha purpurea, A. (Phaenocoma) prolifera Barnesii,* and *A. purpurea grandiflora.* Young plants require to be kept well pinched, to form a good foundation for a large well-furnished specimen. The plants grow well in good fibrous peat and silver sand, with plenty of nodules of charcoal and sandstone sprinkled through it; draining the pots thoroughly, and potting very firm. As the plants grow, the shoots should be neatly trained in the desired form, using as few stakes as possible, and neatly "threading" the intermediate shoots to keep them into their position. The attractive silvery appearance of their stems and leaves, and the bright rosy-purple of their beautiful "everlasting" flowers, make the Aphelexis always very inte-resting objects.

AZALEA.

With the exception, perhaps, of the Erica, there is no tribe of greenhouse flowering plants so useful as the Azalea to the exhibitor. The numerous first-class varieties of *A. indica* which are now in cultivation give the exhibitor a wide choice in selecting the best for his purpose. In flower they make a brilliant and highly attractive display, and are a prominent feature of flower-shows in spring and early summer. A cool greenhouse is their best winter quarters; and when they are wanted for early spring shows, they must be introduced to heat in due time to get them into flower. They bear mild forcing remarkably well, and from six to eight weeks, accord-ing to the variety, is generally sufficient to get them into full bloom. A temperature of 55° at night, rising 5° to 10° in the daytime, is better than harder forcing. Keep plenty of moisture in the atmosphere, and shade from bright sunshine. When the flowers begin to expand, reduce both heat and moisture, and gradually harden them off, so as to enable them to stand exposure at the exhibition. They should be re-potted when needful soon after flowering; using fibry peat, and a little turfy loam and leaf-mould, with a good sprinkling of silver sand and charcoal nodules. Place them in a moist atmosphere, with a heat of 60° to 65° at night, and shade from sunshine till they have made their growth and set their buds, when they should be gradually hardened off, and may be set out of doors in autumn to thoroughly ripen; moving them into winter quarters before there is any danger of frost.

BORONIA.

The Boronias are a useful and elegant genus of greenhouse plants, and form handsome specimens when properly treated.

Among the best for exhibition are *B. Drummondi*, *B. elatior*, *B. hetrophylla*, and *B. pinnata*. They thrive best in fibry peat and loam, with a good sprinkling of sand, and small bits of sandstone and charcoal freely mixed. The pots should be well drained, and the soil made very firm. Young plants should be closely pinched, so as to form good bushy specimens.

BOUVARDIA.

There are many species and varieties of these useful greenhouse plants in cultivation. Bouvardias are not often used as exhibition plants, but when they are well cultivated and trained, they make very handsome and attractive specimens. A few of the best varieties for this purpose are Alfred Neuner, Brilliant, Dazzler, Hogarth, President Garfield, and Vreelandi. They should be potted in a compost of two parts of sandy loam, one of peat, and one of sand and decomposed manure well dried and rubbed down. The best way to grow and train them into specimens is to pinch them well while the plants are young, and keep them near the glass to secure strong, sturdy growth. The shoots should be spread out like those of a trained Erica, and if the flower-trusses stand erect without stakes, so much the better. When the plants begin to flower, they are greatly benefited by liquid manure from sheep-droppings and soot.

CAMELLIA.

For many years the Camellia was the favourite winter and spring flower, and although it has had to give place to a certain extent to the improved forms of the Chrysanthemum, still as a greenhouse flowering plant it has few rivals. The following are twelve of the best: Alba plena, Augustine superba, Bella portuensis, C. M. Hovey, Countess of Orkney, Fimbriata, Imbricata, Jubilee, Lavinia Maggi, Mathotiana, Targioni, and Valtevaredo. Their cultivation is similar to that of the Azalea, except that the major part of the compost should be turfy loam, with a little peat, rotted manure, and bone-meal added. The potting, when necessary, should be done after flowering, and before much growth is made ; but oftener a top-dressing of rich soil is all that is required, especially with large plants. A vinery, if not too much shaded, is an excellent house for them to grow in ; but as soon as the growth is finished the plants should be hardened off, to prepare them for standing in a sheltered place outside. Success for the next year depends upon judicious watering and perfect ripening of the wood during summer and autumn. The buds should be thinned to one or two on each shoot, and when they begin to swell, the plants should be watered with liquid manure until the flowers are fully expanded.

CLIANTHUS.

This is a genus of quick-growing plants, which, although scandant in habit, if well pinched and the shoots regulated, are capable of being grown into neat bushes suitable for exhibition. Two of the best species for the purpose are *C. Dampieri* and *C. puniceum magnificum*. The soil best suited for their growth is a rich fibrous loam and peat in equal parts, with a little sand and leaf-mould added. When the flowers appear, a few doses of clear liquid manure greatly improves the quality of the blooms.

CORREA.

A genus of neat-growing, fine showy plants, of easy culture. Two of the best for exhibition are *C. cardinalis* and *C. magnifica*, the former scarlet, and the latter of a white colour. The soil best suited for them is half light loam, the other half consisting of peat, leaf-mould, and sand. The plants should be slightly pruned after flowering, and as soon as they have started to grow, they should be potted, and after their growth is made, treated the same as Camellia.

CRASSULA.

This genus, known also as Kalosanthes, belongs to a succulent class of plants, with very pretty flowers, mostly of a scarlet colour, which, when freely produced, have a showy and charming effect. *C. coccinea* and *C. jasminea* are very suitable subjects for exhibition, especially as amateurs' plants. They should be potted amongst sandy loam in a lumpy state, with the addition of broken pots and bits of old mortar and charcoal. They do not require much water, especially in winter, but a little liquid manure when they are in full growth gives them a vigorous and healthy appearance, and much improves the flowers. They should be trained with the lower branches in a horizontal position, and the others staked thinly and equally over the whole of the half-globe shape. The pinching of the tops to form side-shoots should be all done at once, so that the heads of flowers which appear the following year may be all developed at the same time.

CYTISUS.

This is a rather extensive genus, the greenhouse species of which are mostly yellow-flowered compact-growing shrubs, and are very useful decorative plants. The following are among the best for exhibition specimens :—*C. canariensis, C. Everestianus, C. filipes,* and *C. racemosus.* They all grow freely from cuttings and seeds, but *C. filipes* is shown to best advantage when grafted on a stronger-growing species, on a stem about three feet high, where, from its graceful drooping habit, it forms a very beautiful specimen. They grow freely in a compost of fibry loam and peat, with a free sprinkling of sand and charcoal. After they have flowered, they should be pruned into shape and repotted. When carefully handled, they form nice pyramidal specimens ; and when the roots fill the pots, they should be liberally supplied with water and occasional doses of liquid manure till growth is finished. They may then be set out of doors to ripen, and be wintered in a cool house, introducing them into a warmer house if necessary to get them into flower for the show.

DAPHNE.

This is a genus of evergreen shrubs, chiefly noted for the sweet perfume of their flowers, and being of a good habit, make, when well grown, fine exhibition plants. Among the best greenhouse kinds are *D. elegantissima* and *D. indica rubra.* They grow freely in a compost of peat, leaf-mould, and sand, and require frequent pinching when young, to get them into a nice bushy form. Good drainage and an ample supply of water while growing are necessary. To secure a profusion of bloom, the main point to be observed is the thorough ripening of the wood by full exposure to light and air, and keeping the plants rather dry.

DRACOPHYLLUM.

The most valuable species of this genus is *D. gracile.* It makes a fine specimen, either trained on a trellis or pinched and neatly staked out in bush form, and should be in every collection for exhibition. It grows best in rough fibrous peat and one-fifth of silver sand. If trained on a trellis, the flowering shoots should be carefully brought to the outside before the flowers begin to open.

EPACRIS.

This is an elegant genus of plants, which is generally associated with the Erica, and although natives of different countries, they thrive under almost the same conditions. A few excellent kinds for exhibition are Hyacinthiflora candidissima, H. fulgens, Miniata splendens, Rubra superba, Sunset, and The Bride. The soil best suited for their culture is rough fibrous peat with a liberal addition of silver sand. After the plants have flowered, the strong-growing kinds should be cut down to within two or three eyes of the old wood, the weaker ones being left a little longer. They should be shifted just after they have started to grow, making the soil very firm, and kept in a close frame till the roots are established in the new soil, after which more air should be given, and the sashes be ultimately removed altogether. It is of much importance in growing the Epacris and Erica that vicissitudes of dryness and heat be guarded against, and therefore they should never be set in an exposed position, where

c

they might suffer from heat and drought. The best mode of training is that which exhibits the greatest quantity of flower with the least degree of stiffness or formality. The shoots should be arching gracefully to the outside, and kept in position by neat stakes and green-coloured threads, which should be removed, if possible, when the plants are exhibited.

ERICA.

As flowering specimens, the Ericas or Cape Heaths may be considered the standard greenhouse plants for exhibition, and when staged in first-class condition, are universally admired, and stand high in cultural merit. To manage them well, they require a house for themselves, or they may be associated with such hard-wooded greenhouse plants as thrive under somewhat similar conditions. They are a very durable race of plants, and by careful management and strict attention to their requirements they can be most successfully grown and exhibited for many years. There are many species and varieties of Erica suitable for exhibition, among which the following hold a prominent place :—*E. Aitoniana Turnbullii*, *E. aristata major*, *E. Cavendishii*, *E. elegans glauca*, *E. Marnockiana*, *E. Massonii*, *E. retorta major*, *E. tricolor Wilsoni*, *E. ventricosa*, *E. v. splendens*, *E. vestita*, and *E. Victoriæ*. The soil they thrive in is good fibry brown peat, torn up into small fragments and the finer particles left out; to this should be added a liberal amount of silver sand, and nodules of charcoal and sandstone, varying the quantity according to the quality of the peat. Drain the pots carefully, and press the soil as firm as possible with the potting-stick, finishing off with a thin layer of fine soil. A judicious and liberal amount of water must be given at the roots at all times, but a more copious supply is necessary during the season of active growth. If the ball of soil once becomes dry, the small fibry rootlets will suffer to such an extent as to cripple the energies of the whole plant, and, if it does not die off in consequence, it will take years to recover itself. The best time to re-pot is when the plants are starting into growth. Large shifts must be avoided; from a six to an eight-inch pot being enough at one time, and possibly for two or three years, all depending upon the progress and health of the plants. Only enough artificial heat should be applied to the house during winter as will keep out frost and dispel damp; and when heat is employed for the latter purpose, it should be applied during the day and in fine weather, with a little air passing through the ventilators.

Mildew is the great enemy of those plants, and when it appears, it must be immediately attended to. A slight dusting with flowers of sulphur over the affected part is the best antidote, and this should be done in autumn over the whole stock as a preventive.

As soon as the plants are done flowering, the old flowers must be carefully removed, and the long shoots pruned or trained into position. In training, it is best to "thread" the shoots into their place with fine but strong green thread, thereby lessening the number of stakes, which injure the roots and are very unsightly when used too freely. This threading is the best method for training the shoots of all such exhibition plants, because when neatly done it is scarcely noticed, and the thread lasts longer than any other tying material.

ERIOSTEMON.

A genus of dwarf evergreen shrubby plants, of a free-flowering nature and of easy culture. The different species flower at various seasons, and make a beautiful display when well cultivated. Two of the best for exhibition are *E. cuspidatum rubrum* and *E. pulchellum*. They succeed in a cool airy greenhouse, and thrive best in a compost of peat and loam in equal proportions, with a good dash of silver sand. Like most hardwooded plants, they must be firmly potted and carefully watered.

GENETYLLIS.

This is a small genus of beautiful plants, natives of Australia, which require careful management to grow them successfully, but are worth some special care, as they make grand exhibition specimens when they are well done. Among the finest are *G. fimbriata* and *G. tulipifera*. They grow best in a mixture of peat and turfy loam, with a free sprinkling of silver sand; draining the pots well to prevent stagnant water from lodging about the roots. They grow well in a cool house along with heaths, and flower in early summer.

FUCHSIA.

When the Fuchsia is in good form and profusely flowered, it is a very striking feature in an exhibition. With the amateur it is a particular favourite, and is perhaps the most frequently seen of all plants in the cottager's window. A few of the best varieties for exhibition are Clipper, Emperor, Eynsford Gem, Lucy Finnis, Phenomenal, and Walter Long. The Fuchsia is very easy to propagate by cuttings, which, with liberal treatment, will form good specimens for exhibition in the second year. After attaining the desired size, the plants may be exhibited for a number of years, if due attention is paid to their management. Starting with a year-old plant, it should be pruned into shape, and placed in heat to break early in March. The temperature of a vinery suits it well at this season. When the growth is about an inch in length, the plant should be shaken out of the pot, the strongest roots pruned, and the plant replaced in the smallest pot the roots will easily fit into, in a mixture of sandy loam and leaf-mould. When the roots reach the side of the pot, the plant should be shifted into one two sizes larger, using a rich compost of fibry loam, well-rotted manure, bone-meal, and wood ashes, with sufficient sharp sand to keep all sweet. As soon as the roots run freely in the new soil, remove the plant to the greenhouse, where, by a judicious application of liquid manure, pinching off the flowers till wanted, stopping the shoots, and careful training, a fine pyramidal plant can be grown during the season, which will form a splendid specimen for exhibition in the autumn.

IMANTOPHYLLUM.

A small genus (known also as *Clivia*) of remarkably handsome plants, and very useful for exhibition. New varieties are being raised, with a richness of colour and beauty which may soon displace the older kinds, and furnish a good variety for selection of this most useful, easily grown, and ornamental class of greenhouse plants. A few of the best varieties of *I. miniatum* are Aurantiacum, Giganteum, General Gordon, Perfection Splendens, and Superbum. They thrive well in a strong rich turfy loam, with a third part of sand and leaf-mould added. After the plants have attained to specimen size, they should not be potted every year, but be top-dressed with a rich mixture of soil and manure. They flower better when pot-bound than when allowed too much root-room. They should be kept in the warmest end of the greenhouse, and be supplied with weak liquid manure when in active growth. The flowering may be forwarded, if required, by placing them in a vinery or forcing-house, as they bear heat and forcing well.

LAPAGERIA.

The well-known and popular *L. rosea* and its white variety are magnificent evergreen climbing plants, and when trained on a balloon trellis make splendid exhibition specimens. To flower them well, they require a light airy position in the house close to the glass, so as to thoroughly mature the wood, for on this, to a great extent, depends the quantity and quality of the flowers. It is a good plan to train the young shoots on strings stretched along the house near the glass, until they complete their growth and show flower, when they should be brought down and regularly trained over the trellis, arranging the flowers that they may all show on the outside of the specimen when they are fully expanded. The plants should be potted in rough turfy peat and loam, with lumps of white sandstone and charcoal, and a sprinkling of bone-meal. The pots should be well drained, as the roots require an abundance of water during the time of growth, and weak liquid manure, given once a week, after the pots are filled with roots, will greatly benefit the plants.

LASIANDRA.

This genus is now generally included under *Pleroma*, but the beautiful free-flowering species known in gardens as *Lasiandra macrantha floribunda* makes such a fine specimen that it deserves special notice, as with proper treatment it is among the very best of flowers, of an attractive rich violet-purple colour. The plant thrives well in light turfy loam with a sprinkling of bone-meal. In such soil, and in a warm greenhouse, it makes a vigorous sturdy growth, which, by careful pinching and regulating, eventually forms a grand specimen. The large flowers are freely produced on the ends of the shoots, and a judicious application of clear liquid manure for a fortnight before they begin to open adds to their size, brilliancy, and endurance. The plants flower best when pot-bound, and when they have attained specimen size they will thrive and flower in profusion for years without shifting, if due attention is given to top-dressing, and feeding them with liquid manure when growth is taking place, as well as when they are coming into flower.

LESCHENAULTIA.

This is a beautiful genus of New Holland plants, the shrubby species of which are of great value for exhibition when they are staged in first-class condition. The two best kinds are *L. biloba major* and *L. formosa.* These plants very much resemble heaths, but being scarcely so hardy, they require more heat in winter, otherwise a heath-house suits them well. Like most of the New Holland plants, they thrive best in sandy peat. They should not be too much confined in the pots, but be shifted on as they require it, taking care not to give too much water when the plants are newly potted. The same mode of training as that recommended for the Heath suits them well.

LILIUM.

This is a genus of bulbous plants, comprising a large number of species, with splendid flowers, all eminently useful for exhibition, and being easily managed, they ought to be more widely cultivated amongst amateurs than is the case. The best species, are *L. auratum* and *L. speciosum*, with their numerous varieties. The soil best suited for their culture is one-half turfy loam, one-fourth peat or leaf-mould and sharp sand, and one-fourth of decayed cow-manure, with some bone-meal added. To secure early specimens, a few pots should be exposed to the influence of sun in the beginning of February, and a little water given to start them. When the shoots are an inch above the soil, the bulbs should be turned out of the pots, and all the old soil picked clear away without injuring the young roots. Clumps of from three to eight bulbs, according to the kind, should be potted in well-drained 12-inch pots, pressing the soil firm about the roots, and leaving a little space for sur-face-dressing. The plants should then be placed close to the glass in a frame where frost cannot reach them, and watered very sparingly till the roots are well through the soil. An abundance of air should be allowed to play amongst the leaves on all favourable occasions, that they may be firm and leathery from the bottom of the shoot upwards, otherwise they will fall off prematurely and weaken and disfigure the plant. After the buds are set, liquid manure from soot and sheep-droppings should be given once a week until the flowers begin to open. The plants may be kept in the greenhouse to bring them in early, or set in a sheltered place outside to keep them back. If they are outside, the pots should be protected from the sun, and a cool moist air maintained around them. The flowering shoots should be supported with neat green-painted stakes, and the leaves kept clean by sponging. After flowering the plants should be gradually dried to ripen them, but the soil should never get dust dry even while the bulbs are at rest.

NERIUM.

This old-fashioned plant, *N. Oleander*, although nearly for-gotten in gardens, may still be seen struggling for existence in some cottager's window or amateur's greenhouse, with long straggling branches and a few small flowers half-way up the shoots, when, by giving a short rest to the plants and judicious pruning and pinching of the growth, with a little warmth when making wood, fine bushy plants with a profusion of bloom would be the result. *N. O. rubrum* and *N. O. album* are the two best for exhibition. Unless the plants are wanted for forcing earlier, they should be pruned in March, and started by watering and placing them in mild heat, such as is afforded to a newly started vinery. After the shoots are several inches long, the plants should be taken to the warm end of the greenhouse, and the growth regulated and pinched if necessary. When the flower-buds appear, all terminal shoots should be pinched, and the plants regularly watered with weak liquid manure once or twice a week. After flowering, the ripening of the wood and resting of the plants are necessary to future success.

PIMELEA.

A genus of shrubby plants, mostly with white flowers, which they produce in great abundance, and are specially attractive as exhibition plants. Two of the best are *P. Hendersonii* and *P. spectabilis rosea*. Being natives of New Holland, the soil that suits them best is fibry peat with a little loam, leaf-mould, and sand. When necessary, they should be shifted in early spring, but large plants continue to do well for several years with a top-dressing only. Although they do not force well, they can, by being subjected to a gentle moist heat, be brought into flower earlier than if grown in the greenhouse. They are the better of a little heat to start them, because it gives longer time to ripen their wood in autumn, which is very important for the success of these plants. They are of a bushy habit of growth, and require very little training, except pinching the strong shoots, and regulating the others to get them into proper form.

PLEROMA.

A genus of very ornamental shrubs, with showy blue and purple flowers and ovate acuminate-shaped leaves of a shining green colour. The best for exhibition is *P. elegans*. The same kind of soil and general culture recommended for Pimelea suits them well, except that they have a straggling habit, and require to be trained into conical shape, to show off the flowers to the best advantage. A large specimen, well furnished with healthy foliage and a profusion of flowers, stands high as an exhibition subject.

PLUMBAGO.

P. capensis and *P. rosea coccinea* are the best exhibition plants of this genus. They are easily managed, and being rapid growers, can be trained into a large specimen in a short time. They thrive well in equal parts sandy loam, peat, and well-decayed manure, with a little silver sand. They should be neatly staked, so that the panicles of flower be all placed in the best position to be well seen, and drooping free and gracefully all over the plant. They should be liberally fed with liquid manure previous to flowering.

POLYGALA.

A genus of showy plants of easy culture. The best of the greenhouse species are *P. acuminata* and *P. grandis*. They thrive best in a mixture of peat and loam in equal parts and some sharp sand. They should be trained in bush form, as they break freely when the growth is pinched; and if supplied with liquid manure when growing, they form dense and profusely flowered bushes.

RHODODENDRON.

This genus of beautiful flowering shrubs is so well known and highly appreciated that its excellent qualities need not be commented on here. Our remarks are almost entirely restricted to the greenhouse kinds or Indian species. Some of the best and freest flowering, including the new hybrids, are Balsaminæflorum and its varieties, Album and Aureum, Duchess of Edinburgh, Jasminiflorum carminatum, M'Nabii, President, Princess Alexandra, Taylori, and Virginalis. The cultivation of Rhododendrons in pots and boxes has to be conducted with great care and consideration ; any mistake in allowing the plants to get dry, which, from the nature of the soil they grow in, is very apt to be the case unless they are closely watched, may ultimately result in failure. They may also be over-watered, which is equally ruinous, but which only occurs through imperfect drainage. The soil they succeed best in is rich fibrous peat with a little leaf-mould, and rough silver sand sufficient to keep the whole open. Large plants do not require to be often re-potted, but they should be annually top-dressed in spring, with the same compost and rich decayed manure.

They should also be watered with weak liquid manure, to assist them in developing their flower-trusses to the greatest perfection. Very little training is needed for large specimens, except tying the shoots to keep them clear of each other, and equalising the space in which they have to grow. Damp on the flowers should be guarded against, and no water allowed on the foliage while the plants are in bloom. The flowering shoots should be well supported, to prevent them shaking when being removed to the exhibition.

Hardy Rhododendrons are sometimes exhibited, and a grand display they make when in full flower. Instead of pot culture for this purpose, the plants should be grown on a peat border specially prepared for them. They should be transplanted every second year to confine the roots, and when wanted, they can be lifted and potted after the buds are set and matured, and afterwards transferred to the greenhouse to bloom.

RHYNCHOSPERMUM.

R. jasminoides is the only species of this genus of importance as an exhibition plant. When it is properly grown and trained into a large and profusely flowered specimen, it has an attractive appearance, and is of considerable weight in competition. It grows freely in a mixture of loam and peat, with a little silver sand and charcoal. It should be trained on a balloon trellis ; but in order to get it into a proper flowering condition, the shoots should be run up thinly on strings close under the glass until the flowers begin to appear, when they should be taken down carefully and trained equally over the trellis, turning the leaves and flowers outwards during the operation. Weak liquid manure should at this time be given twice a week, and continued until the flowers are well expanded.

STATICE.

Amongst greenhouse flowering plants the Statice has few rivals, and a large specimen, well flowered, has a very imposing effect in a collection of exhibition plants. They are best shown to advantage when the flowers are newly developed, as they have a rather faded appearance afterwards. They are of a compact habit of growth, and require a warm greenhouse during winter; but ordinary greenhouse temperature suits them well during summer. *S. Holfordi*, *S. imbricata*, and *S. profusa* are the best for exhibition. They thrive best in turfy loam, leaf-mould, and thoroughly decayed manure, with a sprinkling of sharp sand and bone-meal. During active growth a little weak manure-water is a good stimulant. When large specimens have been secured, they are difficult to handle, and are not easily re-potted, which, however, is not necessary, as the surface soil can be removed to a depth of a few inches, and the space filled up with rich compost, into which the roots will soon run and luxuriate. They are easily raised from cuttings of half-ripened wood, put in about midsummer in sand and fine peat, placing them in mild bottom-heat till they are rooted. The tops of the shoots can also be notched and mossed, the same as directed for Aralias, and when the moss is filled with roots, the tops can be taken off and potted, keeping them shaded till they are growing freely, and they will make nice plants in much less time than cuttings.

TETRATHECA.

This is a favourite genus with exhibitors of hardwood green-house plants, and furnishes some of the most attractive specimens in that class. The best for exhibition are *T. ericæfolia*, *T. hirsuta*, and *T. verticillata*. Like all New Holland plants, they thrive well in a mixture of fibry peat and silver sand, with a free sprinkling of nodules of charcoal and sandstone. They require to be well drained and firmly potted, and, like all this class of fine-rooted plants, they must have an abundant supply of water at the roots while growing, and careful attention to watering at all times, so that they never become dry, which quickly proves fatal to them. The shoots should be carefully regulated, and neatly staked in bush form, " threading " the majority of the finer stems, so that the specimen is not overloaded with stakes,—a very unnecessary and objectionable feature in all hard-wooded specimen plants.

VALLOTA.

V. purpurea, or Cape Lily, is an evergreen bulbous plant from South Africa, which forms a fine flowering specimen for autumn shows. When well flowered it has a brilliant appearance, the rich scarlet flowers forming a fine contrast with the deep green foliage. Select the best bulbs, and pot them before the roots start in rough fibry loam, sharp sand, and dry well-rotted manure. No water should be given till the roots are started, after which they may be set in a cold frame, exposed to the sun with plenty air, and be liberally fed as soon as flower-scapes appear. If wanted in flower at a given time, they can be pushed forward in a mild forcing heat.

JUDGING.

The chief points in judging greenhouse flowering plants are —(1.) Size and freshness of flowers and profusion of blooms; (2.) richness, purity, and substance of petals; and (3.) size, health, and vigour of the plant.

SECTION IV.

GREENHOUSE FOLIAGE PLANTS.

ABUTILON.

A useful genus of ornamental foliage plants, the flowers of which are also highly attractive. The variegated kinds make handsome exhibition specimens when properly treated, the beautifully marked foliage having a very pleasing effect. Among the best for exhibition purposes are *A. Darwinii tesselatum*, *A. Sellovianum marmoratum*, and *A. vexillarium igneum*. They are easily cultivated, and grow freely in a compost of light turfy loam, peat, leaf-mould, and silver sand. The drainage should be perfect, as they delight in a copious supply of moisture at the root while growing, but it should never be allowed to become stagnant. The shoots must be regularly pinched and trained as they grow, to form neat symmetrical specimens. They should be grown in comparatively small pots, applying clear liquid manure to cause them to make free growth and large leaves; and as the growth attains maturity they should be exposed to all the light and air possible to bring out the rich colours of the foliage to the greatest perfection.

ARALIA.

Among the Aralias which thrive well in the greenhouse are several handsome kinds useful to the exhibitor, such as *A. (Oreopanax) dactylifolia*, *A. Sieboldii argentea*, and *A. trifoliata*. They are easily cultivated in a compost of equal parts of turfy loam, peat, and leaf-mould, with a free sprinkling of silver sand and nodules of charcoal. Abundance of water must be given whilst growth is taking place. The finest leaves are put forth in a rather close moist atmosphere, but special care is necessary to prevent the growth being drawn, and air must be given on all favourable occasions. They are easily propagated by notching and mossing the tops of the best furnished shoots.

ARAUCARIA.

An interesting genus of coniferous plants with a symmetrical habit of growth and pleasing appearance. When the specimens are brought out in their best form, they are an effective feature in an exhibition. The best for this purpose are *A. excelsa glauca* and *A. Rulei elegans*. They grow freely in good fibry loam mixed with a little leaf-mould and sharp sand. The coolest part of the greenhouse suits them well, and when they have attained specimen size they should be kept in rather small pots to prevent too rapid growth.

ARUNDO.

A small genus of plants belonging to the order Gramineæ. The variegated species *A. Donax variegata* makes a tall, slender, graceful plant, and when well grown and the stems clothed with fresh and perfectly coloured leaves, it has a stately, airy appearance. A native of Southern Europe where it grows by the water-side, indicates the treatment it should receive when grown in pots. A mixture of fibry loam, peat, and sharp river sand suits it well.

BAMBUSA.

This is a genus of graceful plants, chiefly natives of semi-tropical countries, where they supply the natives with numerous articles of daily use in domestic life and for constructive purposes. The bamboo-cane is the stem of one of the species, and is a valuable plant in the rural industry of the countries where it grows naturally. Among the best for cultivation as exhibition specimens are *B. aurea, B. Fortunei variegata, B. gracilis, B. Simonii*, and *B. striata*, which all grow well under the same treatment as the Arundo. A few applications of liquid manure are beneficial to these plants while they are growing.

BEAUCARNEA.

A genus of plants with gracefully pendant leaves, from two to three feet long. They have a slender stem with a thick swelling at the base, a peculiarity which at once attracts the attention and excites the interest of those who see it for the first time. Among the best species are *B. glauca*, *B. longifolia*, and *B. recurvata*, and all are natives of Mexico. They grow well in rich sandy loam, leaf-mould, sand, and a sprinkling of charcoal. They should be encouraged to grow by placing them in a little heat during the spring months, after which they should be placed in the greenhouse, and supplied with weak liquid manure once a week, and an abundance of clear soft water at all times. They are very slow growers; but if these instructions are carried out, it will be found that they make good exhibition specimens in a comparatively short time.

CLEYERA.

The beautiful-leaved *C. japonica variegata* makes a very handsome plant, and forms a fine exhibition specimen when in first-rate condition, and four to five feet high. It thrives well in light sandy loam and peat, and it should receive an occasional dose of liquid manure when the pot is full of roots. A cool house suits it best.

COBÆA.

The free-growing and attractive greenhouse climber *C. scandens variegata* is the best plant of this genus for making a suitable specimen for exhibition, and when in its best condition and neatly trained on a pyramidal trellis, it has a very fine effect. A mixture of sandy loam and peat in equal parts, with a free sprinkling of sand and leaf-mould, grows it well. The colour of the foliage assumes its finest tints when the plant is a little pot-bound; and when in that state it requires the aid of liquid manure to grow its leaves to their largest dimension and finest colour. It comes true from seed, but the best way to secure fine variegation is to strike cuttings taken off a well-coloured plant.

COPROSMA.

Of this neat-habited genus of pretty-leaved plants, *C. Baueriana variegata* is the only suitable species for cultivating as a foliage plant for exhibition. It is of a shrubby habit of growth and very distinct in the variegation of its leaves. It should be potted in two-thirds loam and one-third of sharp sand and leaf mould. After potting in spring, the plants should receive a little heat till the growth is made, after which the leaves will colour best in a cool greenhouse. They can be trained into a pyramid by frequent pinching; but as a longer time is required to produce a specimen than by tying the shoots on a trellis, the grower can adopt the style of training which best suits his purpose.

CYCAS.

This is a genus of stately plants, very often included amongst palms, although they belong to an entirely distinct family, but possibly the derivation of the name, which is Greek for a palm, accounts for the mistake. Two excellent species for exhibition purposes are *C. circinalis* and the *C. revoluta*. They should be grown in strong loam and sharp sand, taking care not to over-pot them, but rather to feed well when the pots get filled with roots. With the best of treatment they require a number of years to make a specimen; but when they have a stem of three feet or more, with a large crown of dark green healthy leaves they are among the best of foliage plants.

DASYLIRION.

A genus closely allied to Yucca, which it resembles in character. The leaves of some of the species, however, are long, narrow, and pendulous, and they form very graceful and interesting specimens for exhibition. Two of the best species with this habit are *D. acrotrichum gracile,* and *D. glauco-phyllum plumosum.* They should be potted in loam, with some peat, sand, and nodules of charcoal. A judicious application of liquid manure secures the finest foliage, upon which the merit of this class of plants entirely depends.

DRACÆNA.

The best of the greenhouse Dracænas or Cordylines are *D. indivisa Veitchii*, *D. rubra*, and *D. stricta congesta.* These plants, when well grown and furnished with leaves to the base, have a noble and graceful appearance, and are high-class exhibition plants. They should be potted in a mixture of rough turfy loam and peat, with leaf-mould and nodules of charcoal. Plenty of water while growing, and a judicious supply of liquid manure, produces the finest leaves, which colour well with exposure to light and air.

EURYA.

All the species of this genus are evergreen shrubs, of which *E. latifolia variegata*, a Japanese plant with particularly beautiful leaves, is the finest foliage subject for exhibition. It is of a compact habit, and can easily be grown and trained into a handsome specimen. It succeeds well in equal parts of good fibry loam and peat, and grows freely in a cool house.

FICUS.

Several species of this genus which thrive well in the stove may also be included in greenhouse plants, as they succeed well under the treatment, except that they are all the better of a little extra heat while making their growth. The most ornamental and finest for exhibition of this class are *F. Cooperii*, *F. dealbata*, and *F. elastica variegata*. They grow well in rich fibry light loam mixed with peat and sand, and grown in rather small pots. Liquid manure and a free use of the syringe increases the size and beauty of the foliage.

GREVILLEA.

This is a handsome genus of plants, chiefly from New Holland. The best species with ornamental foliage are *G. elegans* and *G. robusta*. They require to be grown in equal parts of turfy loam and peat, with a sprinkling of bone-meal and silver sand.

HOYA.

This is a genus of climbing plants with beautiful waxy flowers. The species have all leaves more or less ornamental, but the only plant with rich coloured foliage is *H. carnosa variegata.* The most suitable soil to grow it well is a thin loam and peat, with a good sprinkling of silver sand. It should be trained on a pyramidal trellis with the leaves freely exposed to the sun, so that the rich colour may be well brought out.

HYDRANGEA.

Although this genus has many species and varieties with splendid heads and panicles of bloom, it has only one variety worthy of a place amongst exhibition foliage plants, viz., *H. hortensis japonica variegata.* The plant should be kept moderately dry, and rested during winter; and, after being started in the spring, it should be partially shaken out and shifted into the same-sized pot, using rather poor loam, a little leaf soil, and sand. It should be kept near to the glass in the coolest part of the house, and supplied with plenty of water during the time of growth. The flower-buds should be pinched out as they appear, and the shoots regulated and trained into position.

LOMATIA.

This is a genus of graceful-habited plants of considerable beauty and value for exhibition. The best of the species are *L. elegantissima, L. ferruginea,* and *L. heterophylla.* They thrive in a compost of light loam, peat, and silver sand, and require an abundance of water while making growth, during which a few doses of liquid manure greatly improves the foliage.

MUSA.

The only species of this valuable genus of tropical plants which may be cultivated in a greenhouse for exhibition specimens are *M. ensete* and *M. superba.* These are two very handsome foliage plants when they are well grown and furnished with an ample head of their handsome leaves. They require a strong rich loamy soil to grow in, with a little extra heat and a copious supply of water when they are making growth. Their large broad leaves, with stout bright crimson mid-ribs, are very effective objects among the other denizens of the greenhouse, and with a little careful management they make grand specimen foliage plants.

PHORMIUM.

A genus of useful as well as highly ornamental foliage plants belonging to New Zealand. The best variegated kinds are *P. Colensoi variegatum*, *P. tenax variegatum*, and *P. t. Veitchii*. They are free growers, and should be potted in strong rich loam and sharp sand, and confined in comparatively small pots, to bring out the variegation in its best shades.

YUCCA.

The variegated forms of this genus are very handsome ornamental foliage plants, and should find a place in every collection. The best of them for exhibition are *Y. aloifolia quadricolor*, *Y. a. variegata*, and *Y. filamentosa variegata*. They grow freely in rich sandy loam and leaf-mould, and succeed very well in an ordinary greenhouse, although they will make specimens quicker in a warmer house. When they have attained a desirable size, they may remain in the same pots for several years, if they receive plenty of water and are regularly fed with liquid manure while they are making growth. At all other times they thrive best when kept rather dry, but not so much as to make them show any signs of flagging ; and they should always be exposed to all the light and air possible, to bring out the colours of the leaves to perfection. They are easily propagated by notching and mossing the tops, and they are also readily increased from the side-shoots, which spring freely from the stem when the top is cut off. These, with a heel to them, strike quickly in sandy soil in a good bottom-heat.

JUDGING.

The chief points in judging foliage plants are : (1.) Healthy, clean, and perfect leaves of a graceful form ; (2.) the colour should be fresh, clear, and well defined ; and (3.) the specimen should be well grown, amply furnished with foliage, and the outline handsome and pleasing.

SECTION V.

ORCHIDS.

The Natural Order Orchidaceæ is a very extensive one, comprising many genera with numerous species and varieties, representatives of which are met with in nearly every part of the world. The flowers produced by some of the tropical species are unsurpassed for brilliancy and beautiful colouring, and the scientific interest possessed by others claims for them a place in public favour. A fine display of orchids at an exhibition is always a point of great attraction to the visitors as well as to the connoisseur.

In selecting orchids for exhibition, special attention should be given to those of a free-flowering nature, with large showy flowers, rich in attractive colours, of a pleasing odour, and having an imposing appearance when the plant is in full bloom. In order to grow the various classes of orchids to perfection, it is considered necessary to have three houses, or three divisions in a range of houses, set apart for them, and usually known as the "East-Indian," "Mexican," and "Peruvian" house, respectively. These houses should be constructed and arranged specially to meet the wants of orchids requiring either tropical, temperate, or cool treatment. The ventilation and heating must be fully equal to all possible demands, and an abundant supply of soft clear water must always be at command. The heating apparatus should be capable of supplying a night temperature of 70° to the East-Indian house, 60° to the Mexican house, and 50° to the Peruvian house, in all conditions of weather. The houses should be fitted in their interiors so that the plants can always be grown near the glass, as they mostly require all the light they can get on dull days, and the necessary shade from bright sunshine can always be given with proper appliances. When shade is required, it is best applied by moveable scrim canvas, of the proper thickness to give the desired amount of shade, working on a roller with pulley and rope, so that it may be run down or up at pleasure, as the state of the weather renders necessary.

The cultural requirements of orchids demand careful study and considerable experience to enable the grower to have them in perfection at a given date for exhibition. With a knowledge of the conditions under which they thrive in their natural habitats, and the length of time they require to make and ripen their growth, and the period at which they have to be started to have them in full flower by a certain date, an exhibitor will be on the right road to success, especially as many species are easily forced if late, and as easily retarded if too early. The packing of orchids for transport to an exhibition requires special attention, and the utmost care should be taken to support every flower-stem and every flower by proper stakes standing well over the spikes, which are securely tied to them, using a layer of wadding round each stake and flower-stem, and tying with soft flat raffia. The flowers should also be kept in position, and safe from being rubbed in transit, by pads of wadding placed around and between them, and secured by a tie when necessary for safety. When the specimens are set on the exhibition table, every scrap of packing material—stakes, wadding, and ties—must be removed, so that the specimens may appear as natural as possible.

AËRIDES.

A large genus of epiphytal orchids, the name of which is derived from the fact that most of the species obtain their nourishment from a moisture-laden atmosphere. Many of the species are very handsome plants, freely producing beautiful spikes of richly coloured flowers, and forming most useful specimens to the exhibitor. Among the finest species for exhibition are A. affine, A. Fieldingi, and A. Lobbii, all noted for their long drooping spikes of delightfully fragrant flowers, which, although not so showy as the flowers of some other genera, have a very telling effect in a competitive collection.

Specimen plants are best grown in perforated pots, three

parts filled with clean-washed potsherds and lumpy charcoal; using live sphagnum only for placing a few of the lower roots in, leaving the majority of them to ramble free in the air. The old material should be very carefully washed out from the tangled roots in a tub of soft tepid water, and then the plant set on a layer of moss pressed firm on the top of the crocks, keeping the base of the stem well above the rim, and working the sphagnum gently but firmly in among the roots. Finish off with a neatly placed layer of live sphagnum, and secure the shoots to stakes which have been fixed firmly among the drainage as it was filled in. The material being used in a moist state, no water will be required at the roots until the plant begins to grow, but the atmosphere must be constantly kept saturated with moisture till the plants have expanded their flowers and finished growing, giving water copiously at the roots while growth is taking place. Care must, however, be taken not to allow water to fall on the flowers after they begin to expand, as it is liable to cause irreparable mischief. From the end of September to early in April a drier and more airy atmosphere should be maintained, and less water given at the roots, just sufficient to prevent any shrivelling of the leaves. As a rule, the months of March and April are the best for shifting all epiphytal orchids, Aërides included ; and as the latter are natives of tropical regions in the Old World, they thrive best in the East-Indian house.

ANGRÆCUM.

This is a beautiful genus of interesting epiphytal orchids, the characteristic feature of which is the peculiar long tail-like spur depending from the flower. The most esteemed for exhibition are *A. eburneum*, *A. Ellisii*, and *A. sesquipedale*. These are all natives of Madagascar, and consequently require to be grown in the tropical or East-Indian house, and the treatment described for Aërides suits them admirably. In their seasons they are highly effective specimens for exhibition, and the curious long drooping " tails " of the beautiful flowers are always an object of great attraction to the public.

CALANTHE.

A useful genus of terrestrial orchids, comprising some very handsome species, which in their respective seasons make a fine display, and are great favourites with the public. The winter-flowering *C. Veitchii* and *C. restita* make very effective specimens for exhibition at that dull period of the year; and *C. Masuca* and *C. veratrifolia* are among the best of the summer-flowering species. Naturally, they are slow in forming exhibition specimens, and are usually "made up" by placing together the required number of plants in a large pot or tub—a practice not to be commended, although with many species of orchids it may be difficult to detect, and in the case of deciduous species, such as Calanthes, Pleiones, and the like, it may be the least of two defects, and as such allowed to pass. Calanthes are easily cultivated, and should be grown in a rich mixture of loam and leaf-mould, potted firmly in well-drained pots of the ordinary type, keeping the bulbs about level with the rim of the pot. They grow best in the Indian house, and while growth is taking place they should receive copious supplies of water at the root, with a supply of weak liquid manure once or twice a week when the pots get full of roots. The deciduous kinds require to be kept dry after they have flowered till they are thoroughly rested, and the best time to re-pot and start them is just when they naturally begin to show signs of growth in spring.

CATTLEYA.

Perhaps no genus of orchids stands so high in popular favour at the present day as that of Cattleya. Most of the species are of easy culture, and under proper treatment soon grow into specimens suitable for exhibition. The gorgeous display of flowers produced by many of the choicest kinds renders them indispensable to the exhibitor, and the magnificence of the blooms always attracts crowds of admirers amongst the visitors to a flower-show. Among the finest for exhibition are *C. Dowiana, C. Exoniensis, C. Lawrenceana, C. Mendelii, C.*

Mossiæ, and *C. Warneri.* These all thrive best when grown in pots or baskets half-filled with clean-washed crocks, over which is placed a layer of moss, and the remainder filled with a mixture of fibry peat and pieces of charcoal, freely sprinkled with silver sand. Keep the plant high over the rim of the pot, and finish off the surface with a neat layer of live sphagnum. They require an abundant supply of water while growing, and a moist atmosphere; but they should be kept in a drier state after the growth is matured, as they grow more freely and flower better after they have enjoyed a long period of rest. Being chiefly natives of Brazil and the adjacent countries, they thrive best in a temperature between that of the Indian and Mexican houses, or what is known as the " Brazilian " or Cattleya house; but they are very accommodating so long as they receive plenty of heat and moisture while making their growth, and, when subjected to a low temperature, careful attention is paid to their wants, and especially not to over-water them.

CŒLOGYNE.

A genus of very showy and useful epiphytal orchids, chiefly natives of the moister parts of the East Indies, and thriving well under the same treatment as Cattleyas. The finest for exhibition are *C. asperata, C. cristata,* and *C. Massangeana.* They delight in abundance of moisture while growing, and should be deluged with clear soft water overhead from the rose of a watering-pot in the growing season. Being evergreens, they should never be allowed, even in the resting season, to become quite dry at the root. Generally they enjoy plenty of heat when making growth, but *C. cristata* is an exception, as it makes the finest growth in a temperate house, and expands its flowers to the greatest perfection in a tropical heat. As soon as the plants are in full bloom, it and most of the Cœlogynes stand well and last longest in a cool temperature. The same compost as for the Cattleyas suits well, and a sprinkling of fish-manure, or any other stimulant of the same nature, over the surface of the soil once a-week while the plants are growing is very beneficial.

CYMBIDIUM.

A few species of this genus form very handsome and effective specimens for exhibition, especially *C. eburneum* and *C. Lowianum*. When these are set up in their best condition, they are exceedingly telling plants in a collection. They thrive best in roomy well-drained pots, in a compost of rough fibry peat, loam, sphagnum, and silver sand. Abundance of water and a weekly application of clear liquid manure suits them well. They grow to perfection in the Cattleya house. They should be thoroughly rested in a cool temperature, but the roots must never be allowed to become dust-dry.

CYPRIPEDIUM.

This is a very beautiful and interesting genus of terrestrial orchids, representatives of which are found in all climates, from Tropical to almost Arctic regions, and furnish the exhibitor with some of the most valued specimens. Among the finest of the tropical species for exhibition are the following :— *C. caudatum*, *C. Harrisianum*, *C. Lawrenceanum*, *C. Sedenii*, *C. Stonei*, and *C. superbiens*. They are easily cultivated, and delight in a good compost of turfy loam, peat, and silver sand, with a little chopped sphagnum and nodules of charcoal. The plant should be potted rather firmly, and kept about level with the rim of the pot, finishing off the surface with a fine layer of live sphagnum. Plenty of heat and moisture should be given while the plants are growing, and less water applied when they are resting ; but they must not be allowed to get quite dry at the root at any season. An intermediate house suits them well during the resting period.

DENDROBIUM.

This is an extensive genus of orchids, comprising many species and varieties, most of which are exceedingly floriferous ; and although the flowers are not so large as those of many other genera, yet their numbers make up for what they lack in size. The colours of some of them are dense, bright, and well marked, and tinged with various shades of red, white, yellow, and purple. Among the best for exhibition are *D. densiflorum*, *D. Devonianum*, *D. Falconeri*, *D. fimbriatum oculatum*, *D. formosum*, *D. nobile*, *D. thyrsiflorum*, and *D. Wardianum*. The three periods of development peculiar to orchids is well defined in this genus : first, the splendid display of flowers ; second, rampant growths; and finally, gradual in-action. During the latter period, the process of storing, con-solidating, and preparing for the future display is steadily proceeding, and the object of the cultivator should be to supply the conditions marked by Nature, and assist the maturation of the plant as far as possible. They should be potted in the same manner and with the same material as is recommended for Cattleya, but a little more brown peat may be used with advantage. When they are in active growth, plenty of clear water and a little liquid manure should be given ; and the plants syringed twice a day. When growth is finished, they should be removed to a cool house with a night temperature of 50°, a dry atmosphere, and plenty of light, the amount of water being gradually reduced, until the shoots are thoroughly matured, after which less water is required.

EPIDENDRUM.

This extensive genus contains many species of little value to the exhibitor. There are some, however, of considerable merit on account of their compact habit and the beauty and durability of their flowers, which entitle them to be classed among exhibition plants. Three of the best are *E. evectum*, *E. paniculatum*, and *E. vitellinum majus*. The first two have rather tall stems, while the last is a low, compact-growing species, sending up from the top of the bulbs its fine showy flowers, which stand for months when placed in a cool shady place. Their culture is similar to that recommended for Cattleya, except that an intermediate temperature suits them best.

LÆLIA.

A genus of orchids comprising several species of a high standard as exhibition plants, such as *L. anceps*, *L. elegans*, and *L. purpurata*, and their splendid varieties. They are natives of the warmer parts of America, from Mexico to Brazil, and require the same treatment as the Cattleya, to which genus they are closely allied. They thrive best in rough fibry peat and sphagnum, mixed with crocks and bits of charcoal, filling the pot about half full of drainage, and keeping the plant high above the rim. They require abundance of water with a little liquid manure during the growing period, and a fair supply of water at all times.

LYCASTE.

Some of the species and varieties of this genus are of great value, not only to the exhibitor, but also from a commercial point of view. They have flowers of much beauty and substance, and the only point against them is that they do not produce so many blooms, nor show them so well, as many others of less importance. Among the best are *L. candida*, *L. gigantea*, and *L. Skinneri*. They are easily cultivated, provided due attention is paid to watering, which should not be altogether withheld when resting ; but a copious supply must be given during the time of growth. They thrive best in peat with a few pieces of old fern roots chopped down and mixed with it. The Mexican house suits them well.

MASDEVALLIA.

An extensive, interesting, and useful genus of orchids from the high mountainous regions of Central America, extending from Mexico to Peru. There are numerous species and varieties, some of which have large, richly coloured flowers, rendering them most valuable as exhibition specimens, while others have curiously marked and grotesquely shaped flowers, of great interest to the botanist and to the collector of rare and curious plants. Those with the largest flowers and bright showy colours are the best for exhibition, such as *M. Harryana*, *M. ignea*, *M. Lindeni*, *M. Schlimii*, *M. tovarensis*, and *M. Veitchiana*, with their varieties. Under good management the plants will flower twice, and some of them thrice during a season, and are very useful for both exhibition and decoration. The cool house is the proper place in which to grow them in, and with due attention to their wants they are very amenable to good culture. A house with a north aspect suits them admirably, as they delight in a cool moist atmosphere, and shaded from bright sun. They require ample drainage, and thrive best in turfy loam, silver sand, and sphagnum, with plenty of water given to the roots while the plants are growing.

MAXILLARIA.

This is a genus of orchids mostly natives of South America. They very much resemble Lycaste in their flowers and habit growth, but have the disadvantage as exhibition plants of concealing their blooms among their leaves; nevertheless, some of the species are of considerable merit on account of the beauty and fragrance of their flowers, and should be in every collection. Some of the best for exhibition are *M. grandiflora*, *M. Harrisonii*, and *M. venusta*. They should be potted in the same material as Lycaste, and kept in the cool house, and, like most orchids, they require plenty of water and a moist atmosphere while growing—receiving, when resting, just enough water to keep the bulbs plump.

MILTONIA.

This is another South American genus, most of the species growing naturally in the moist warm parts of Brazil. They are low-growing, compact, evergreen plants, and as they throw up one or two flowers from their small but numerous bulbs, there are generally plenty of flowers on a specimen to produce a fine display. Some of the best for exhibition are *M. candida*, *M. spectabilis*, and *M. vexillaria*. They should be potted in brown peat and moss, with plenty of charcoal and crocks, and treated afterwards similar to Cattleyas.

ODONTOGLOSSUM.

This is one of the most beautiful and deservedly popular genera of the whole family, and is of the easiest culture; also, when the plants are in flower, they can be used for decorative purposes with the smallest risk of damage. The graceful spikes of bloom when cut can scarcely be equalled for all kinds of floral decorations, and the variety of form, colour, and marking is not excelled by any other genus. They are natives of various parts of Central America, and are often found growing at high and cool altitudes. Many of them make handsome exhibition specimens when large and well flowered. Some of the best, including their choicest varieties, are *O. cirrhosum*, *O. crispum*, *O. grande*, *O. Insleayi*, *O. nebulosum*, *O. Phalænopsis*, and *O. Pescatorei*. They thrive best in a compost of brown fibry peat with a little sphagnum and charcoal, and the whole neatly covered over with live sphagnum kept fresh and moist. With the exception of a few species, a cool house is most suitable for their cultivation, and they require plenty of water at all seasons, as they do not go to rest even in winter. The species requiring a higher temperature than what is generally given, such as *Phalænopsis*, should be gradually dried off during winter, and given a plentiful supply of water after they show the flower-spikes and while growing. They should be closely shaded from strong sun, and a little weak liquid manure from soot and sheep-droppings is a good stimulant when in active growth.

ONCIDIUM.

This is a most extensive and handsome genus, many of the species producing long, branching, and numerously flowered spikes, which make them very striking and useful as exhibition plants. Some of the best, inclusive of their choice varieties, are *O. ampliatum*, *O. crispum*, *O. macranthum*, *O. splendidum*, *O. tigrinum*, and *O. varicosum*. They inhabit the tropical parts of America, including Brazil, Mexico, and the West Indies. Many of the species thrive best in baskets or on blocks of wood, but the stronger growing kinds do best in pots. The weaker growing should have the conditions of pot culture made as near an imitation as possible of a basket. They grow well in perforated pots, in a compost consisting of fibry peat, sphagnum, and pieces of the roots of the common bracken chopped small. The kinds above mentioned should all be grown in the Cattleya house, receiving plenty of moisture, which should be moderated when the plants are at rest.

PHAIUS.

This is one of the terrestrial genera of orchids, of tall slender growth. The species are spread over a wide range of country, being found in Asia, India, and Australia. They are noble-looking plants, with long plicate leaves and tall spikes of beautiful flowers. Two of the best for exhibition are *P. grandifolius* and *P. Wallichii.* In common with all the terrestrial class of orchids, they require a substantial compost to grow them well, consisting of three parts turfy loam, from which the small particles have been sifted, and one part of sharp sand and well-decayed manure. The heat of the East-Indian house, with plenty of water, suits well while they are growing, but they should be removed to a cooler house and given very little water when resting. They are greatly benefited by a liberal supply of weak liquid manure when they are in active growth.

PHALÆNOPSIS.

A splendid genus of orchids, which, however, require time and good management to produce first-rate specimens; indeed, they are among the most difficult to grow satisfactorily of all the Orchid family. The leaves of some of the species are of great substance, large, and prettily spotted, but they are rather liable to rot at the base. For exhibition, the best are *P. amabilis, P. grandiflora,* and *P. Schilleriana.* They are natives of the warmest parts of India and the Malay Archipelago, and are very sensitive to cold and damp during the resting season. The warmest end of the Indian house is best suited for their growth, where a moist atmosphere should be provided while the plants are growing, but kept moderately dry when they are resting. They should be very carefully potted, using the best fibry peat and sphagnum, with a good sprinkling of charcoal and potsherds, keeping the roots well above the rim of the pot, and care taken to shade them from bright sunshine.

SACCOLABIUM.

A very fine genus of epiphytal orchids from India and the Malay Archipelago, and when well grown they furnish excellent exhibition specimens. The best for this purpose are *S. ampullaceum*, *S. Blumei majus*, and *S. violaceum*. They are very similar to Aërides in their habits, and require the same treatment in their cultivation. They should be grown near the glass, so as to receive all the light possible, and only be shaded slightly when they are growing.

SOBRALIA.

This is a genus of tall slender-growing orchids, with large showy flowers produced on the top of the stem. The flowers are rather short-lived, and fade in a few days, but they succeed each other so rapidly that there are always sufficient on a large specimen to make a fine display. The best for an exhibition specimen is *S. macrantha*. It thrives in the Mexican house, and should be potted in a mixture of rough peat, chopped sphagnum, potsherds, and charcoal. It requires a good supply of water, with a little weak liquid manure while growing, and to be kept moderately dry when at rest.

STANHOPEA.

This is a very peculiar genus of orchids from Tropical America, with thick, waxy, grotesque-formed, variously coloured flowers. Instead of the flower-scape shooting upwards, it takes the opposite direction, which renders the plants unsuitable for pot culture. They do best grown in baskets, but they can be grown in pots with fair success if the pseudo-bulbs are raised well above the rim, and a strict watch kept on the scapes as they begin to push, so as to guide them to the outside. Some of the best for exhibition are *S. grandiflora*, *S. insignis*, *S. Martiana*, and *S. tigrina*.

They thrive well in sphagnum alone, but a little fibrous peat amongst it is beneficial to their growth. The East-Indian house suits them well, and they should receive plenty of water while growing, with a long season of rest, during which time they should not be over-watered.

TRICHOPILA.

This is a pretty genus of dwarf evergreen orchids, natives of Central America. They have curiously formed flowers, which proceed from the rhizome at the base of the bulb, some-times drooping over the edge of the pot. They are very showy, and should be tied up to thin stakes, and well exposed. Some of the best for exhibition are *T. crispa*, *T. fragrans*, and *T. suavis*. They require to be potted among brown peat with a little sphagnum and charcoal, and the bulbs should be well elevated above the pot, so that the scapes may extend freely. They should be kept near the glass in the Mexican house, and sparingly watered at all times.

VANDA.

An exceedingly handsome and free-flowering genus of orchids, from the same regions in the tropics as Aërides and Saccola-bium, and thriving well under similar treatment in the Indian house. They are admirably adapted for exhibition, and take a high place in competition. A few of the best of them are *V. Batemanii*, *V. insignis*, *V. suavis*, and *V. tricolor*, the three latter comprising each several very beautiful varieties, which are popular among cultivators of orchids. Vandas luxuriate in a high and moist temperature while they are growing, with a copious supply of water at the roots, and an occasional appli-cation of clear liquid manure, the air being kept saturated by a free use of the syringe and damping of the paths and stages. As the growth reaches maturity the plants should be kept rather drier, and receive all the light and air possible, while at the same time the high temperature of the house is main-tained, and no more shading applied than will prevent scorch-ing of the foliage. The plants thrive best when grown in baskets, which should be nearly filled with potsherds and rough lumps of charcoal, with a layer of sphagnum laid over them, on which the roots are placed, and filling in among them a mixture of chopped sphagnum, potsherds, sandstone, and charcoal. The material should be used more or less rough according to the size of the plant and capacity of the basket, rounding it well up over the lip of the basket, and finishing off with a neatly placed layer of live sphagnum.

ZYGOPETALUM.

A genus of handsome epiphytal orchids from Tropical America. Most of the species are tall-growing plants, sending up terminal scapes bearing racemes of showy flowers. They bloom during the winter, which makes them of value as decorative plants, although not so useful for exhibition. Their blooming period can, however, be slightly retarded by keeping the plants in the cool end of the Cattleya house, and by introducing them to a cool house when in bloom they will stand fresh for months. Some of the best species are Z. *Mackayi*, Z. *maxillare*, and Z. *rostratum*. They should be grown in well-drained pots, among fibrous peat and pieces of charcoal, which should be covered over with live sphagnum. They should have plenty of water when growing. but be kept moderately dry when resting.

JUDGING.

The chief points of merit in exhibition orchids are : (1.) The abundance and quality of the flowers, which should be fully developed, fresh, large of their kind, of good substance, pure and bright in colour, and the markings clearly defined ; and (2.) the size and vigour of the plant, a *bona fide* specimen formed of a single plant taking precedence of a " specimen " made up of several plants, all other points being equal. In orchids, however, so much depends on the particular variety, and the conditions under which it is exhibited, that the knowledge and experience of the judge can alone decide their relative merits.

E

SECTION VI.

FERNS.

This is the most graceful section of cultivated plants, and always forms a highly interesting class at horticultural exhibitions, never failing to attract the attention and excite the admiration of all classes of the people. It comprises numerous genera and a multitude of species and varieties of lovely aspect, some of which are to be found in every habitable part of the world, from the stately wide-spreading tree-fern of the tropics to the lowly spleenwort of the northern Highland glen. As a rule, they are shade-loving plants, and delight to rear their fragile forms under the protection of taller and more robust subjects. There they are safe from the blast and the scorching rays of the sun, and enjoy the requisite calm and moisture to build up their delicate structure without interruption.

Following the usual practice at flower-shows, we will divide this Section into five Sub-Sections, namely :

(1.) Exotic Ferns.
(2.) Hardy or British Ferns.
(3.) Filmy Ferns.
(4.) Tree-Ferns.
(5.) Selaginellas.

1. EXOTIC FERNS.

Among the best for exhibition of the numerous, handsome, graceful, and useful ferns in this Sub-Section are the following :—

ADIANTUM.

This is one of the best known and probably the most extensively grown genus of the whole family. Among the best for exhibition are *A. cardiochlœna, A. cuneatum, A. Farleyense, A. Flemingii, A. formosum, A. gracillimum, A. Sanctœ Catharinœ,* and *A. tenerum.* These are all strong growers, and form excellent specimens when well managed. They thrive best in a shady place in the stove, with a strong heat and moist atmosphere. They should be grown in loam and sand, with some charcoal and pieces of brick broken small. They should receive plenty of water while growing, with weak liquid manure once a week when the pots are full of roots. When cutting out the old fronds in spring, some of the freshest should be left to support the young ones, and to assist in promoting stronger growth.

ASPLENIUM.

This is one of the most interesting genera of ferns, comprising a host of species and varieties, which are spread over most of the habitable globe. Among so many it is not easy to select the best, but the following suit admirably for exhibition :—*A. Belangerii*, *A. dimorphum*, *A. formosum*, *A. ferulaceum*, and *A. viviparum*. Both the stove and greenhouse species thrive well in a compost of loam and peat in equal parts, with a liberal addition of silver sand. When the plants have attained to specimen size, they should not be potted every year, but top-dressed with rich mould, and liberally fed with weak liquid manure when they are in active growth. This practice should commend itself to all growers of ferns for exhibition, because the plant in all its parts must be grown to the fullest possible dimensions, which cannot be done if the roots are starved ; and, with the exception of one or two genera, the whole family are much benefited by the judicious application of liquid manure while they are growing.

BLECHNUM.

The species of this genus are found, although in limited numbers, in all parts of the world. They naturally grow about the sides of streams and in crevices of rocks, where they get plenty of shade and moisture. The length of their fronds range from six inches to as many feet, and are generally of graceful shape. Among the best for exhibition are *B.*

corcovadense, B. lanceolatum, and *B. longifolium gracile.* The first two thrive best in the cool end of the stove, while the last should be grown in the greenhouse. They should be grown in a mixture of peat, loam, and sandstone.

DAVALLIA.

All the species and varieties of this useful genus form beautiful specimens, and are of great service to the exhibitor. Among the finest for that purpose are *D. canariensis, D. elegans, D. Fijiensis plumosa, D. hirta cristata, D. Mooreana,* and *D. pyxidata.* The stove kinds require a strong moist heat and a plentiful supply of water at the roots, and should be grown in baskets or well-drained pans, using a compost of fibrous peat and sharp sand, with a large proportion of charcoal and sandstone, to allow the water to pass freely away. The greenhouse kinds should be grown among firmer material, with less charcoal in it, and receive less water at the roots.

GLEICHENIA.

This is a specially attractive race of ferns, and indispensable to the exhibitor. Among the best for exhibition are *G. circinata glauca, G. dicarpa longipinnata, G. dichotoma, G. flabellata, G. Mendelii,* and *G. rupestris.* They all require greenhouse treatment, and should be grown in well-drained pans or shallow tubs, using a compost of fibry peat and silver sand with a little charcoal. They grow naturally on the banks of streams where the water overflows their roots at times, thus indicating that they require a plentiful supply of moisture. They require to be firmly potted, leaving a hard even surface for the wiry creeping stems to run upon. This genus of beautiful ferns have rather delicate and easily injured roots, and liquid manure or manure in any form should not be given to them. They can be increased by division or by layering the stems in small pots.

GYMNOGRAMMA.

This genus includes the most beautiful of the "golden" and "silver" ferns. They stand in the front rank of exhibition ferns, and no collection is complete without them. Some of the best "golden" ferns are G. *chrysophylla* and its varieties, *Laucheana* and *Parsonsii* (crested); and of the "silver," *G. peruviana argyrophylla*, *G. tartarea*, and *G. Wettenhalliana* (crested). They all belong to the stove section, and require a strong moist heat for the development of their fronds. With good management they soon grow into large specimens, and abundantly compensate for the attention given them. They thrive best when potted in a compost of sandy loam and peat, with some bits of charcoal. They should be well supplied with water at the roots during the growing season, and a little liquid manure once a week; but all syringing must be dispensed with, as it washes off the powdery "gold" and "silver" from the fronds, and spoils the plant for exhibition.

LASTREA.

This is a very extensive genus, containing many species and varieties, which, with few exceptions, belong to temperate climes, and thrive well in the greenhouse. Some of the best for exhibition are L. *aristata variegata*, L. *elegans*, L. *lepida*, L. *Richardsii multifida*, and L. *tenericaulis cristata*. These are all of easy culture, and grow freely in equal parts of turfy loam and peat, with a sprinkling of silver sand.

LOMARIA.

A genus of useful ferns, containing several species of merit as exhibition plants; and although some of them inhabit the warmest parts of both hemispheres, yet very few require stove-heat. A few of the best are L. *discolor*, L. *falcata*, L. *gibba*, and L. *Patersonii elongata*, all of which thrive well in the greenhouse. They grow well in sandy loam and peat, and are all the better to be started and grown for a time in a warm moist atmosphere, taking care to gradually inure them to the temperature of the greenhouse.

LYGODIUM.

A genus of climbing ferns which grow rapidly, and can be trained into any form. Two excellent species are *L. palmatum* and *L. scandens*. The first is a native of North America, and grows well in a greenhouse ; the last belongs to the East Indies, and requires to be grown in the cool end of the stove. Light loam, leaf-mould, and silver sand suits them admirably, and weak liquid manure given once a week when the soil is full of roots is beneficial. The long stems are best shown when trained in a spiral manner round a balloon wire trellis, being careful not to break off the points of the shoots till the top is reached. Large specimens trained in this fashion, with strong healthy fronds, are among the finest exhibition ferns.

NEPHRODIUM.

A very numerous and useful genus of ferns, natives of many parts of the world, some of which form handsome specimens for exhibition. A few of the best are *N. cuspidatum, N. deltoideum, N. molle corymbiferum, N. patens cristata,* and *N. villosum.* These thrive well in an intermediate temperature, and should be grown in fibry loam, leaf-mould, and sand, with plenty of moisture at the roots, and a little liquid manure once a week in the growing season.

NEPHROLEPIS.

This is a genus of graceful and ornamental ferns, well adapted for exhibition. The species are all stove plants, but a few thrive well in a warm greenhouse. Some of the best are *N. cordata, N. davallioides furcans, N. pectinata,* and *N. pluma.* They grow well and form fine specimens with the same treatment as recommended for Lomaria.

NOTHOCHLÆNA.

A beautiful and interesting genus of stove ferns, having the same characteristics as the gold and silver Gymnogrammas, and thrive well under the same treatment. Some of the best for exhibition are *N. chrysophylla, N. lanuginosa, N. nivea, N. sinuata,* and *N. trichomanoides.*

OSMUNDA.

A genus of large handsome ferns, mostly natives of North America, Asia, and Japan, but one is the well-known Royal Fern, *O. regalis*, of the British Isles. They possess fronds varying in length from one to three feet or more, and when well cultivated form fine specimens for exhibition. Three of the best are *O. cinnamomea*, *O. javanica*, and *O. palustris*. They are all moderately hardy, and thrive remarkably well in the greenhouse. The ordinary fern soil suits them, and as they grow naturally in damp marshy situations, they should be kept moist even when dormant.

PLATYCERIUM.

A remarkably distinct genus of ferns, containing a few species which, when well grown, possess a unique and handsome appearance, and form grand specimens for exhibition. Some of the best are *P. alcicorne majus*, *P. grande*, and *P. Hilli*. These are all tropical species, and require similar treatment to other stove ferns, except that rough material should be used in potting and the plants set high in the pot, which should be well drained. They are better to be kept rather dry at the root during winter.

POLYPODIUM.

This is a very extensive genus, widely spread throughout the world, giving a choice of hardy, medium, and tender species. The following are some of the best exhibition kinds:—*P. appendiculatum*, *P. Krameri*, *P. loriceum*, and *P. nigrescens*. The first and last require stove treatment, the other two a greenhouse temperature. If well cultivated under ordinary conditions, they readily form good exhibition specimens.

POLYSTICHUM.

A very ornamental genus of rather hardy ferns, except a few of the species which require stove heat. The following three are greenhouse species, and suitable for exhibition :—

P. flexum, *P. mucronatum*, and *P. setosum*. They are very easily cultivated, and if started in a little heat in spring, make excellent exhibition specimens during the summer.

PTERIS.

This is an ornamental and very useful genus, of the easiest culture. Among the best species for exhibition are *P. argyræa*, *P. cretica albo-lineata*, *P. serrulata cristata*, and *P. tricolor*. They may either be grown in the stove or greenhouse, but make the finest plants in a little heat when proper care is taken to keep them clean and free from scale, from which they are liable to suffer if neglected. Sandy loam and peat with a little maltster's "cummings" grows them well. This material is the refuse from malt-kilns, and when mixed with soil for ferns and soft-wooded plants it has a very beneficial effect upon them.

WOODWARDIA.

The finest species of this genus is the beautiful *W. radicans cristata*, which makes a fine shapely plant when grown in a warm greenhouse. It grows well in a suspended basket, and thrives best in equal parts sandy loam and fibry peat.

2. BRITISH FERNS.

This Sub-Section includes all ferns indigenous to the British Isles. They should be exhibited in two classes, viz., large and dwarf specimens, the latter including all kinds with fronds less than one foot in height from the surface of the pot. British ferns generally require the same cultural treatment, and the few exceptions to the rule will be mentioned under their own genus. They should be grown in moderate-sized pots, well drained; and they thrive well in peat, loam, and sand, varying the proportions of these according to the delicate nature or robust habit of the species. They grow well in the open air, but to produce them in the best condition for exhibition, they should be grown in a glass-house or pit with a north aspect. They require plenty water at the roots, and a moist atmosphere during growth; but they should be kept moderately dry during winter, and protected from extreme frost, if possible, without using artificial heat. A long period of undisturbed rest is a sure preliminary to a vigorous start and luxuriant growth.

ADIANTUM (*Maidenhair*).

The beautiful " Maidenhair fern," *A. capillus-Veneris*, is the only British species of this genus. The fronds grow from 8 to 12 inches in length, and when it is started and grown for a time in a warm moist place, it forms a splendid dwarf exhibition specimen.

ALLOSORUS (*Parsley Fern*).

A very pretty little fern is *A. crispus*, the only species of the genus indigenous to Britain. It grows naturally in rocky places on mountain-sides, well out of the reach of water, which it seems to avoid; and when cultivated in a pot it should be well drained, and the plant never watered overhead, nor kept in an atmosphere over-charged with moisture. It makes a first-rate dwarf exhibition specimen when in its best condition.

ASPLENIUM (*Spleenwort*).

There are several species of this genus natives of the British Isles, some of which are rare, while others are quite common. Some of the best for pot culture and exhibition are *A. Adiantum-nigrum*, *A. lanceolatum*, *A. marinum*, *A. Trichomanes cristata*, and *A. viride*. These are all dwarf ferns, but with good treatment some of them will attain dimensions to qualify them for the large specimen class.

ATHYRIUM (*Lady Fern*).

A. Filix-fœmina is the only species of this genus which is a native of Britain. It is a very common plant in every part of the United Kingdom, generally growing in damp sheltered situations, where it often attains a height of five feet or more,

and forms one of the noblest specimens of British ferns. Some
of the best varieties of it are *A. F.-f. Frizelliæ*, *A. F.-f. latifolia*,
A. F.-f. multifidum, and *A. F.-f. Victoriæ*.

BLECHNUM (*Hard Fern*).

The hardy erect-habited *B. spicant* is the only British
species of this genus, and is found on heaths and in woods
all over the country. Three of the most suitable varieties for
exhibition are *B. s. multifidum*, *B. s. ramosum*, and *B. s. triner-
vum*. They are all dwarf ferns, and make very pretty plants
when well grown.

CETERACH (*Scale Fern*).

The somewhat rare British fern *C. officinarum* is found
generally in rocky places and on old ruins. It is rather
difficult to cultivate, and should be grown in well-drained pots
in a mixture of fibrous loam, pieces of sandstone, charcoal,
and old mortar, and set in a shady corner of the house, where
it should be carefully watered and the crown kept thoroughly
dry. It is one of the smallest of ferns, but its peculiar appearance makes it very interesting and unique as an exhibition
specimen.

LASTREA (*Buckler Fern*).

This genus contains several British species, with numerous
varieties, which are widely scattered over the country. Among
the best for large specimens are *L. cristata*, *L. dilatata lepi-
dota*, *L. Filix-mas cristata*, and *L. spinulosa*. These are all
easily grown, and produce fronds varying from two to five feet
in length—the crested varieties in particular forming grand
specimens.

OSMUNDA (*Royal Fern*).

This genus has one native species only, from which has been
obtained a handsome crested variety. *O. regalis* and *O. regalis
cristata* are two of the finest for exhibition among British
ferns. The former attains to a height of four or five feet,
and when well furnished has few rivals, while the latter fills
a prominent place in the most choice collections.

POLYPODIUM (*Polypody*).

This genus includes the favourite Beech, *P. Phegopteris*, and Oak, *P. Dryopteris*, Ferns, beside the numerous beautiful varieties of the common Polypody, which are among the best for pot culture. They are all excellent for exhibition, the Beech and Oak Ferns making fine dwarf specimens, while the Polypody, with good treatment, grows to a fair size as a specimen.

POLYSTICHUM (*Shield Fern*).

This is one of the larger-growing genera of British ferns, the species and varieties of which are very useful and ornamental, and form splendid specimens for exhibition. Some of the best are *P. aculeatum proliferum*, *P. angulare cristatum*, *P. a. plumosum*, and *P. a. proliferum*.

SCOLOPENDRIUM (*Hart's-tongue Fern*).

There is only one British species of this genus, but there is a great choice of varieties. They are mostly all pretty evergreens and very easily cultivated. Some of the best for exhibition are *S. vulgare crispum*, *S. v. cristatum*, and *S. v. ramosum majus*. They form nice dwarf exhibition plants, and contrast well with others.

WOODSIA (*The Woodsia*).

These are among the smallest and most interesting of British ferns, of which there are only two species, *W. hyperborea* and *W. ilvensis*. They grow naturally on rocky mountains, but are amenable to pot culture, and form beautiful dwarf specimens for exhibition. The pots should be extra well drained and the soil never allowed to become dry.

3. FILMY FERNS.

With the exception of Todea, which assumes a considerable size, the genera included in this Sub-Section are of a lowly and unassuming growth, but the peculiarly delicate, almost transparent, nature of their fronds, and their extremely neat and graceful appearance, render them specially attractive and interesting in an exhibition, at which they ought to be more often seen. The best known genera, *Hymenophyllum*, *Todea*, and *Trichomanes*, are all very similar in their nature and habits, and thrive well with the same treatment. Where possible, a small house with a north aspect should be specially devoted to them, and where that cannot be afforded, they should be grown in cases or frames inside of other houses; and as they are shade-loving plants, the case may be placed anywhere under the shade of other plants, so that a moderate share of light and air reaches it. A close, moist, moderately cool atmosphere, kept closely shaded from sunshine, suits them well; and most of them thrive in well-drained pots, in a compost of fibry peat, chopped sphagnum, and silver sand, freely mixed with chips of sandstone and charcoal, used more or less rough according to the size and nature of the plants. The air must be kept saturated with moisture by frequent sprinklings with tepid water; and the drainage must be perfect to allow it to escape, as they will not bear stagnation at the roots. Some of the creeping species thrive best on blocks of wood or pieces of tree-fern stems, with a little sphagnum tied on, to retain moisture for the roots as they run over the surface of the block.

HYMENOPHYLLUM.

The numerous members of this genus are mostly natives of tropical regions, but a few beautiful species are found in New Zealand. The best for exhibition are *H. ciliatum*, *H. flexuosum*, *H. hirsutum*, and *H. scabrum*. *H. Tunbridgense* and *H. Wilsonii*, which are natives of Britain, are occasionally exhibited in collections of hardy ferns. They are generally found in sheltered crevices of rocks with the water from above trickling down among their roots, and keeping the air heavily laden with moisture, giving an idea of what to aim at in growing them to perfection.

TODEA.

This is by far the most effective genus for exhibition among the filmy ferns. Many of the species form elegant plants, but require many years to make a large specimen. Some of the best are *T. Fraseri*, *T. pellucida*, *T. plumosa*, and *T. superba*. They develop much larger fronds when placed in a mild heat during the growing season, when the conditions as to moisture and shade are maintained.

TRICHOMANES.

A highly interesting genus of filmy ferns, including a large number of beautiful species with clear membranous fronds, varying in length from an inch to about a foot. The great majority of them are natives of tropical countries, but one lovely species, the well-known Killarney Fern, *T. radicans*, is a native of Britain. They are rather easier to cultivate than the other two genera, but they thrive well under the same general treatment. Trichomanes grow best in shallow pans, with the compost raised well above the rim. A little turfy loam and plenty of lumps of sandstone added to the compost is beneficial. Beside *T. radicans*, *T. angustatum*, *T. maximum*, *T. reniforme*, *T. rigidum*, and *T. venosum* are all useful and interesting plants for exhibition.

4. TREE FERNS.

These are a stately and attractive class of plants in an exhibition, and are frequently employed, along with palms of some height, to break up the smooth formality of the arrangement of other exhibits of lower stature, and as a set-off to the brilliant colours of flowering plants, with a very pleasing effect. Tree-ferns require to be grown in rather lofty houses, where they have ample room to develop to the fullest extent their large and graceful fronds. To assist them in the production of fronds of the largest size and most graceful form, it is a good practice to cover the stem with a thin layer of sphagnum, kept in its place by fine string of the same colour. The sphagnum should be applied in spring before growth commences, and it should be kept saturated by daily pouring a copious supply of water on to the top of it at the base of the fronds, and plying the syringe freely over it several times a day. Fresh roots will soon begin to push from the crown, and, running down the moist sphagnum, will give increased vigour to the fronds. Tree-ferns delight in abundance of moisture in the air while growing; but care must be taken to have the fronds well matured, in a more airy and drier atmosphere, before the date of the exhibition, so that they may stand exposure and not be damaged in transit. Owing to the pliable nature of their fronds, tree-ferns can be covered up and taken to a flower show with less trouble than almost any other class of plants; still, all due precaution should be taken not to injure a single frondlet which would in the least detract from the merits of a perfect specimen.

Although tree-ferns are a numerous class in many tropical and sub-tropical countries, the number of genera in cultivation is very limited, and not more than three or four of those generally grown in our stoves and conservatories are ever seen at exhibitions in tree-like proportions, with clear stems of six feet or more in height. The best of these are included in the following list.

ALSOPHILA.

A very handsome genus of tree-ferns, producing fronds of magnificent proportions under good treatment, which have a telling effect in any plant arrangements. The finest species for exhibition are *A. australis* and *A. excelsa*, both of which thrive best in a warm greenhouse, although they are hardy enough to grow well in any glass structure from which frost is excluded. They grow freely in a mixture of turfy loam and peat, plentifully sprinkled with small bits of sandstone and charcoal. They must have a copious supply of water while growing, and a fair allowance at all times, and the young fronds must be shaded from the direct sunshine till they are full-grown. The stem of Alsophila is generally rather slender, but by judicious mossing it can be increased to any desirable girth, and with advantage to the size of the head of fronds.

BLECHNUM.

It is seldom Blechnums are exhibited as tree-ferns, and unless their stems exceed three feet clear in height, they should be excluded. When over that height, and furnished with fresh and well-developed heads of fronds, they are very effective in a group of plants. The only species generally grown which assumes a useful tree-like habit is *B. braziliense*, and its two varieties, *B. corcovadense* and *B. crispum*.

BRAINEA.

Only one species, *B. insignis*, is in cultivation, a very beautiful and interesting tree-fern from Hong-Kong, which forms a highly effective plant in a collection when it is shown in its best condition. In this country it is seldom seen with a stem of sufficient height to be exhibited in the tree-fern class. It thrives well in an intermediate house, in a comparatively small pot, filled with a compost of fibry peat and loam, with a liberal sprinkling of silver sand and charcoal nodules. It requires abundance of water and shade while making its growth.

CYATHEA.

A genus of tree-ferns, comprising some of the handsomest species in cultivation, and of great usefulness for either decorative or exhibition purposes. Among the best are *C. Burkei*, *C. dealbata*, the handsome and popular silver tree-fern from New Zealand; *C. insignis*, known in gardens as *Cibotium princeps*, and *C. medullaris*, one of the stateliest of all the New Zealand species. *C. insignis* is a native of the West Indies, and thrives best in a stove, but the others succeed admirably in a greenhouse, kept a little warm and moist when they are pushing out their new fronds. They all grow well in the usual compost for tree-ferns, — fibry loam and peat, with a free sprinkling of sand and charcoal nodules, and plenty of drainage, in rather small pots. Abundance of water and shade from sunshine while growing, and a moderate amount of both at all other times suits them well.

DICKSONIA.

The Dicksonias are a princely race of tree-ferns, the genus including the greatest number of handsome arborescent species of any grown in this country, and furnishing the exhibitor with his principal specimens in the tree-fern class. Among the finest for exhibition is the handsome and well-known favourite *D. antarctica*, which is oftener seen in good form at horticultural shows than any other tree-fern; *D. Berteroana*, an elegant and rather rare species from the Island of Juan Fernandez; *D. regalis* (*Cibotium regale*), a graceful species from Mexico; and *D. squarrosa* from New Zealand, one of the most esteemed of all for exhibition when in first-rate condition. All the species mentioned thrive well in a cool greenhouse, but make the finest heads of fronds when they receive the assistance of a little extra heat and abundance of moisture while growing. A copious supply of soft water, of the same temperature as the air of the house, should be daily poured over their crowns from the time the fronds begin to shoot up till they are fully expanded. Like all tree-ferns, Dicksonias should not be over-potted, and may remain in the same pot for years if they are regularly attended to otherwise and the stems kept moist in the growing season. After the fronds have attained their full size, they remain longest fresh and green in a moderately dry and cool atmosphere shaded from bright sunshine. Light fibry loam and peat, with plenty of nodules of sandstone and charcoal, firmly pressed into the pot or tub, grows them to perfection.

LOMARIA.

Several species of this useful genus of ferns assume an arborescent habit with age, but few of them ever attain sufficient stature to entitle them to be exhibited as tree-ferns. *L. discolor*, *L. gibba*, *L. Magellanica*, and *L. zamiæfolia* are the best of the erect-stemmed species; and when they reach three feet or more of a clear stem, surmounted by a well-balanced head of their graceful fronds, they are really beautiful objects, and of much service to the decorator as well as to the exhibitor. They thrive best in a temperate house, and when intended for tree-ferns they should be grown in as small pots as possible, filled with the usual fern-growing soil,—light fibry loam, peat, and sand; kept shaded while growing, and receive abundance of water, especially overhead, till the fronds have reached their full size.

TODEA.

A very interesting genus of lovely ferns, of which *T. Fraseri* and *T. superba* are the best for exhibition of those that form arborescent stems. *T. superba*, when in its finest form, is perhaps the most graceful of all ferns, and is a most valuable exhibition plant. It is seldom possessed of a stem of the necessary height to entitle it to compete in the tree-fern class, where, at its best, it is liable to be dwarfed by the stately Dicksonias, Cyatheas, and Alsophilas; but with a well-grown head of perfect fronds three feet or more across, it is generally a more meritorious plant than its bulkier neighbours. Todeas thrive well in a cool temperature, but they grow fastest and lengthen their fronds more rapidly when grown in a close moist stove. They delight at all times in a close humid atmosphere, well shaded, and the plants kept moist by frequent sprinklings overhead with soft, clear, tepid water. A compost of fibry peat, silver sand, and nodules of charcoal, all well mixed and firmly pressed into pots one-third filled with drainage, grows them well.

5. SELAGINELLA.

This is an extensive and singular genus of plants, very much diversified in appearance. Some kinds are short, thick, close, and as flat as a carpet, while others have long trailing stems reaching out several feet from the root. A dozen of the best kinds for exhibition are *S. africana, S. apoda, S. caulescens, S. erythropus, S. grandis, S. hæmatodes, S. Kraussiana, S. Lyallii, S. Martensii, S. uncinata, S. Wallichii,* and *S. Wildenovii*. With the exception of *Martensii* and *Wildenovii*, which grow well in a greenhouse, the others require a stove or an intermediate temperature and a moist atmosphere to grow them well. They should be planted in well-drained pans in fibry peat, leaf-mould, silver sand, and small chips of charcoal and sandstone, with some turfy loam added to the compost for the stronger-growing kinds. For the production of large exhibition specimens, various modes of growing and training have to be resorted to. The low-growing, dense,

F

carpet-like kinds should be grown nearly flat, with a slight rise to the centre. This is accomplished by stretching rings of wire around the outside of the pan, placing over them a piece of wire-netting, covering it with moss, on which the soil is laid, and the broken-up pieces of the plant with roots thickly planted all over it. The straggling growers should be trained as pyramids, or in the form of an obtuse cone. This is done by widening the base, as already described. A light hollow-framed cone may be placed in the centre, to reduce the dead weight of the specimen and the quantity of material required. Then begin at the bottom and build up around the cone layer upon layer of light turfy peat or loam, or pieces of both, laid flat and with a regular outward surface, introducing at short intervals a little fine soil with rooted pieces of the plant until the apex is reached. A well-grown specimen of this description, about four feet high, is a great attraction at an exhibition.

JUDGING.

All ferns should be of graceful habit and pleasing outline, and be set up for competition in a tasteful manner, so as to show the merits of every specimen to the best advantage.

The general points of merit in ferns are: (1.) Fronds well developed, and perfect in every part; (2.) graceful or stately in habit; and (3.) the colour of the fronds fresh and bright.

Selaginellas should be (1.) well grown, fresh, and of good colour, without speck or blemish; (2.) the free-growing kinds graceful and well balanced; and (3.) the "carpet" varieties, with a short, close, even pile.

Tree-ferns should not be admitted for competition in that class with less than three feet of a clear stem, measured from the surface of the soil to the base of the crown of fronds. The points of merit are: (1.) Round shapely stem, at least three feet in length; (2.) a well-developed graceful head of fronds, without blemish; and (3.) fresh and clear in the colour. All points being equal, the taller the stem, with a proportionate girth, the better the specimen.

SECTION VII.

TABLE PLANTS.

This is generally a well filled and keenly contested class at horticultural exhibitions, and attracts a considerable share of attention from the public. The exhibits are usually confined to small well-grown specimens of fine-foliaged plants, because they are the most popular and useful section for table decoration, their easy culture, durability, and graceful appearance rendering them peculiarly suitable for the purpose. The other sections which are occasionally exhibited at flower-shows are flowering plants, berried plants, conifers, and ferns, all, however, being eligible for exhibition in the one class for "Table Plants," unless specially divided in the schedule. Of foliage plants there is plenty of choice, but the numbers of suitable species in the other sections are rather limited, and not more than six of them, if confined to *distinct species*, should be required in competition. As a rule, it is better to stipulate for *distinct varieties* only, and if the merits of the collections are equal, then the number of species will settle the point, the greatest number receiving the highest award.

Among the best of the first or Foliage section of Table Plants, the following genera, with a single species or variety characteristic of the most suitable type, may be selected :—*Acalypha musaica, Aralia Veitchii gracillima, Croton Chelsoni, Dracæna Sidneyi, Eulalia japonica variegata, Grevillea elegans, Lomatia elegantissima, Maranta Veitchii;* of Palms, *Cocos Weddelliana* and *Phœnix rupicola* are two of the best representatives of the class; *Pavetta borbonica, Phrynium variegatum, Rhopala elegantissima,* and *Terminalia elegans.* The culture of this section is of the easiest, and the proper soil and general treatment is the same as that already described for specimen foliage plants. Vigorous young plants must be secured, and they must be grown as quickly as possible to maintain them in robust health to the desired size. Rich soil, good drainage, firm potting in small pots, and a liberal but judicious use of artificial and liquid manure, combined with a free application of the syringe and perfect cleanliness at all times, will grow them to perfection.

The second section, or Flowering Plants, is the most attractive, including such subjects as *Amaryllis, Amasonia punicea, Anthurium, Azalea, Begonia, Cyclamen, Dielytra, Epiphyllum, Eucharis, Fuchsia, Gesnera,* Orchids, *Pancratium, Primula, Poinsettia, Rhododendron, Spiræa, Thyrsacanthus,* and *Vallota.* As a rule, flowering plants are much more difficult to grow to perfection and exhibit in full beauty at a fixed date than foliage plants, but their generally elegant habits and the beauty of their flowers are worthy of special efforts being made to obtain these plants in perfection. Their general culture is described in other sections, and the special treatment required to produce handsome, and well-flowered specimens is nearly similar during their growing stage to that recommended for foliage plants. Healthy young plants to start with, liberal treatment, attention to watering and cleanliness, and a good position near the light when nearing the flowering stage, to form and develop the flowers, is about the surest road to success.

The third section, or Berried Plants, are a very limited class if confined to plants bearing what are popularly known as "berries;" but if plants with showy fruits are admissible when produced on specimens of a proper size, the range of subjects is considerably widened. Among the best are *Ardisia, Aucuba, Callicarpa, Capsicum, Dianella,* Hardy Fruit Trees and Bushes, Oranges, *Rivina, Skimmia,* and *Solanum.* To have the plants in this section in their best condition, much care is necessary, and to succeed no trouble or watching must be grudged.

Of the fourth section, Ferns, there are numerous species of a graceful habit suitable for competition as table plants. Among the best are *Adiantum, Asplenium, Athyrium, Blechnum, Lastrea, Lomaria, Neottopteris, Nephrodium, Osmunda, Pteris,* and *Todea.* There are only certain species and varieties of the genera mentioned that are suitable for table plants, those with graceful fronds about one foot in length being most eligible for the purpose. The pinnæ of certain Adiantums, such as *A. Farleyense*, being too heavy for the slender footstalks, should be supported on fine galvanised wire with a hook at the top to hold the frond in its natural position.

The fifth section, Conifers; this, the last section of table plants, is, with few exceptions, the best for exhibition, their hardiness and durability being greatly in their favour. For competition, they should be grown under glass, and they well repay a little trouble to bring out their best colours and points of merit. The following are some of the best :—*Araucaria excelsa, Arthrotaxus selaginoides, Biota (Thuia) orientalis aurea, Cupressus Lawsoniana albo-variegata, Juniperus chinensis aureo-variegata, Libocedrus chilensis, Retinospora plumosa albo-picta, Thuiopsis dolabrata variegata, Thuia occidentalis elegantissima,* and *Wellingtonia gigantea.* All of these are grown from cuttings in the usual manner, but care should be taken to select the most vigorous cuttings from the points of sideshoots, so that they may soon form well-balanced young plants. Sound fibry loam of a medium texture suits them well, and they should be potted rather firm in well-drained pots. They delight in a free circulation of air, a moderate supply of water at all times, and all the light possible to bring out their best gold and silver variegation. In summer, syringe every afternoon to keep them clean and fresh.

JUDGING.

The points of merit in a table plant are : (1.) Gracefulness of habit, with every twig and leaf fresh and complete; (2.) the colour of flowers, foliage, and berries should be bright, without blemish, and the markings clear and well defined, pure white, bright red, and green taking precedence of yellow and blue. No table plants should be exhibited in pots over six inches in measurement, and clean-stemmed plants are preferable to those of a squat or bushy habit with many stems or footstalks rising from the surface of the soil. The height of a table plant should be about 20 inches including the pot, which should have a suitable cover over it, and the surface of the soil neatly covered with fresh green moss.

SECTION VIII.

PALMS.

The order of Palmaceæ is confined very much to tropical regions, where many of the species grow with a stately grandeur and luxuriance which we can never hope to see them attain under cultivation in our hothouses. Still, many of them are grown with much success in our stoves and greenhouses, especially those of lower stature, which accommodate themselves better in the limited space that can be afforded for their development. They are all noble-looking plants when they have attained to a fair size, and many of them are extremely graceful even while in a young and small state. It is among these that the best and most useful kinds for exhibition are found, a selection of which is given in this chapter, with a few notes on their characteristics and treatment. As a rule, palms are best when raised from seed, which in general is the only method of increasing them.

ARECA.

A very graceful genus of palms, most of which are tropical plants and require to be grown in the stove, although they stand well for weeks at a time in conservatories and moderately heated halls and rooms. The best for exhibition includes *A. aurea*, *A. lutescens*, and *A. rubra*. These grow rapidly under good treatment, and soon attain a considerable size. They grow well in a compost of fibry loam, peat, leaf-mould, and sand ; and when they attain specimen size, loam with a sprinkling of bone-meal suits them best. A moist atmosphere while growing and plenty of water at the roots is necessary. Liquid manure should be given when the pots are full of roots.

CALAMUS.

An elegant genus of slender, rapid-growing palms, chiefly from the East Indies, and consequently they require a strong moist heat with plenty of water during the growing season.

C. asperrimus, *C. ciliaris*, and *C. plumosus* form fine graceful specimens in a few years, and are the best for exhibition. They thrive well in a mixture of loam with a third of leaf-mould and sand. They throw up suckers freely from the roots, which can be taken off and potted to increase the stock.

CARYOTA.

A handsome genus of palms with bipinnate leaves from six to eight feet in length. Some of the species are rather hardy and dwarf growing, while others grow to a great height. *C. Cummingii* is a very elegant dwarf species, while *C. Rumphiana* is tall and strong growing. *C. sobolifera* is a very graceful, slender-growing dwarf palm, which throws up plenty of suckers, by which it can be readily increased. They require to be firmly potted in soil similar to that recommended for Calamus.

CHAMÆDOREA.

This is a very elegant dwarf genus of palm, including numerous useful species, some of which are particularly hardy, which is a good recommendation for exhibition plants. Three of the best species are *C. elegans*, *C. graminifolia*, and *C. Wendlandii*. They thrive best in a light open peat with a little turfy loam and sand, and they require plenty of moisture and close shade while making their growth.

CHAMÆROPS.

This is a genus of fan-leaved palms, natives of temperate latitudes, and comparatively hardy, so that they are well adapted for greenhouse culture. The best for exhibition are *C. excelsa* and *C. humilis*. They are very easily cultivated, and grow well in rather strong loam, with some leaf-mould, sand, and small pieces of charcoal. They can be increased from suckers, which they throw up abundantly after they have reached some size.

COCOS.

Some of the species of this genus are the finest for exhibition in the whole family. The species *C. Weddelliana* from

South America is one of the most graceful palms grown, and is always admired, whether exhibited as a table plant or as a specimen. It thrives well in one-half lumpy fibrous loam, and the other half peat, leaf-mould, and broken pieces of sandstone. It enjoys strong heat and a moist atmosphere to grow in, and must be kept free from insects.

CORYPHA.

A small genus, which in the form and size of its foliage is a striking contrast to the Cocos. Some of the species have the most magnificent fan-shaped leaves of any in the noble class to which they belong. *C. australis*, a greenhouse, and *C. umbraculifera*, a stove species, are the best for exhibition. They require good drainage and abundance of water, and grow well in a compost of half-fibry loam, the other half peat and sand.

ELAËIS.

A small genus of useful palms, one species of which produces that valuable commodity the palm-oil of commerce. Two of the best for exhibition are *E. guineensis* and *E. melanococa*. They are most useful decorative plants while small or half specimen size. They should be grown in rich loam with a few pieces of sandstone or charcoal to keep it open, and luxuriate in heat and moisture while they are growing.

GEONOMA.

This genus shares pre-eminence with Cocos for beauty and gracefulness. Most of the species are dwarf and slender in the stem, which gives them a light and elegant appearance. Among the best for exhibition are *G. gracilis*, *G. Martiana* and *G. procumbens*. They require to be potted in light spongy peat, fibry loam, and leaf-mould, with a little sand and nodules of charcoal. They are almost aquatics, and much more water should be given them than is generally supplied to palms. To maintain the dark-green colour, they should be closely shaded when growing, and a slight shade afterwards.

KENTIA.

A robust, very ornamental, and rather hardy genus, well suited for an intermediate house, and are all excellent subjects for exhibition specimens. Some of the best are *K. australis, K. Belmoreana, K. canterburyana,* and *K. Fosteriana.* They thrive best in a mixture of loam and peat in equal parts, with silver sand and nodules of charcoal. Unlike many other palms, they grow best when not too much confined at the roots, and they must be copiously supplied with water at both roots and overhead. Liquid manure when they are growing has a beneficial effect.

LATANIA.

This is another genus of fan-leaved palms, all the species of which possess that massive and majestic appearance peculiar to the tribe, and are of the utmost importance as exhibition specimens. Among the best are *L. borbonica, L. glaucophylla,* and *L. rubra.* They should be grown in strong loam with a third of leaf-mould and a sprinkling of half-inch bones and nodules of sandstone. They luxuriate in a moist hot stove while making growth, but stand well in a cooler and drier house when the foliage is matured, and they are exceedingly useful as decorative plants.

LICUALA.

A genus which belongs to the fan-leaved section, and inhabits the moist, warm parts of India. They generally possess a slender stem, and require a long time to form a specimen of any height. Among the finest are *L. elegans, L. grandis,* and *L. peltata.* They grow well in peat and loam in equal parts, with small pieces of sandstone and charcoal. They should receive an abundance of water both at the roots and overhead while growing, and an occasional dose of liquid manure when the pots are full of roots.

LIVISTONA.

This is a noble genus, some of the species growing in their native habitat to the height of a hundred feet.

They are strong robust growers, and eminently adapted for exhibition purposes. Three of the best are *L. altissima*, *L. Jenkinsiana*, and *L. rotundifolia*. They grow freely in equal parts loam and peat, with some sharp sand and half-inch bones. When pot-bound, a judicious application of top-dressing and liquid manure greatly improves their appearance.

OREODOXA.

Another small genus of tall-growing palms, including the famed cabbage-palm, *O. oleracea*, of the West Indies, with long pinnate arching leaves and slender stems, much swollen at the base. Two of the best species are *O. oleracea*, *O. regina*, and *O. Sancona*. A mixture of loam and peat with a little sand is a suitable compost, and they luxuriate in heat and moisture while growing.

PHŒNICOPHORIUM.

The handsome *P. seychellarum* is a rare and striking palm from the Seychelles Islands, with leaves about eight feet long and four feet broad when at their best. Their margins are curiously cut up into segments, which look like a deep vandyke fringe, and have a very fine effect. It thrives well in rough peat and fibry loam, with some broken sandstone. It requires a high moist temperature and a copious supply of moisture in the air and at its roots to fully develop the magnificent leaves, which frequent doses of liquid manure greatly help.

PHŒNIX.

A genus of useful and ornamental palms, of which the best and most suitable for exhibition are *P. dactylifera* (date-palm), *P. reclinata*, and *P. rupicola*. To grow them well, they require to be potted in equal parts of loam and peat, with plenty of heat and moisture, and a free use of liquid manure.

PRITCHARDIA.

This is a small genus of recent introduction, from the South Sea and Pacific Islands. The best exhibition species are *P. filamentosa* and *P. pacifica*. They should be grown in fibry peat and loam in equal parts, and a little sand and charcoal. They require a plentiful supply of water during the growing season, and a fair allowance at all other times, and are much benefited by a little clear liquid manure.

PTYCHOSPERMA.

This is a handsome genus, with slender, smooth stems and gracefully arching pinnate leaves, which have a very pretty effect. The young leaves assume a reddish tinge of colour before they are fully expanded, after which they turn dark-green. This peculiarity of the species is very striking and beautiful. The best are *P. Alexandræ* and *P. rupicola*. They delight in a compost of two-thirds loam and one of peat and sharp sand, with a high temperature and abundance of moisture while they are growing.

RAPHIS.

A genus of fan-leaved palms which are very suitable for exhibition specimens. The best species are *R. flabelliformis*, with its variegated form, and *R. humilis*. They thrive well in fibry peat and loam in equal parts, with a little sharp sand added. They should receive liquid manure while growing, which assists in developing the foliage and bringing out the colour to perfection.

SEAFORTHIA.

The two species of this genus in cultivation, *S. elegans* and *S. robusta*, are both elegant and useful palms. They are comparatively hardy, and make good exhibition specimens. Strong loam, leaf-mould, and sand make a good compost for them, and they thrive well under the usual treatment.

STEVENSONIA.

This genus is composed of only one species, *S. grandiflora*, one of the noblest of all the palms in cultivation. It is a grand foliage plant, and forms a fine exhibition specimen. A native of the Seychelles, it requires treatment similar to what is recommended for Phœnicophorium.

THRINAX.

A genus of the fan-palm, some of the species of which are very elegant and well adapted for exhibition. Three of the best are *T. elegans*, *T. multiflora*, and *T. parviflora*. They should be grown in a mixture of rich loam, peat, and river sand, and luxuriate in a warm moist atmosphere while making their growth.

VERSCHAFFELTIA.

This genus includes one grand species for exhibition purposes, *V. splendida*, which is a very distinct palm, and forms a fine contrast among others of the family. It is a native of the Seychelles Islands, and requires similar treatment to the Phœnicophorium.

GENERAL REMARKS.

The general details of the cultivation of every species of palm are given at length in many excellent works on stove and greenhouse plants, easy of access to all wishing for information, so that we have confined our remarks to a few leading points in their special culture for exhibition. To that we may now add some useful hints of general application to the class. Every precaution should be taken, and every means used, to preserve the foliage of palms in a perfect state, as upon that alone are they valued as exhibition specimens. Insects of every kind should be diligently watched for, and all known means of prevention used to keep the plants free from their attack; and should they ever appear, no time should be lost in extirpating them. Palms delight in a rather close moist atmosphere while they are making growth, and are much benefited by a judicious use of artificial and liquid manures. As a rule, they thrive best when grown in moderate-sized pots, which the roots can quickly fill after a shift. Great care must always be exercised in watering newly shifted palms, because over-watering is certain to sour the soil, and the result will be disappointment; at the same time, they must not be allowed to suffer from want of water, the result of which will be equally detrimental to the health of the plant. They should not be subject to any sudden changes, and when nearing the show-day, they ought to be carefully and gradually hardened off, so as to be able to bear the exposure without risk of injury. In preparing the plants for transport to the exhibition, the leaves should be carefully drawn up towards the centre, and be securely tied to a stout stake, round which cloth and cotton-wool has been wrapped, to prevent injury by rubbing. The pinnæ should be all neatly laid in, and secured with strands of flat raffia, using tissue-paper and cotton-wool wads wherever necessary, to prevent crushing or rubbing. When all are securely laid in, the whole should be wrapped in a soft cloth or sheet, and the plant may then be transported any reasonable distance in perfect safety, even in coldish weather if frost does not prevail. All packing materials must be entirely removed from every specimen before it is set up for competition; and after they are placed, a careful glance around ought to be given, to see that every leaf is shaken out free, and all in perfect order. Equal care should be given to the repacking and taking home of exhibition plants, because every blemish to the foliage is permanent, and can never be got rid of till the leaf is removed. When safely returned to their quarters, they should be kept a little close and moist for a few days, when they will pick themselves up, and look as fresh and beautiful as if they had never been out of the house at a flower-show; and whenever they are wanted, they are ready again to do the same work, which, in short, means "successful exhibiting."

JUDGING.

In judging Palms the chief points of merit are: (1.) A free graceful habit; (2.) size and freshness of the leaves; and (3.) size, health, and cleanness of the plant.

SECTION IX.

MISCELLANEOUS EXHIBITION PLANTS.

Among the leading features in this highly important section of exhibition plants are : (1.) The autumn and winter flowering Chrysanthemum ; (2.) bulbous plants in flower, the best of which are the Amaryllis, Hyacinth, Lily of the Valley, Narcissus, and Tulip ; (3.) soft-wooded plants, which flower naturally in winter and spring, or are amenable to mild forcing, such as the Auricula, Balsam, Begonia, Calceolaria, Cineraria, Coxcomb, Cyclamen, Geranium, Gloxinia, Hydrangea, Mignonette, Pelargonium, Petunia, and Primula ; and (4.) hardy plants, which force easily, and produce a rich display of the most beautiful colours when so treated, including Azalea, Deutzia, Dielytra, Lilac, Rhododendron, Rose, Spiræa, and many others.   In the prize schedules of the leading Horticultural Societies most of the kinds enumerated are favoured with one or more classes to themselves, and they also form an important feature in all collections and groups of plants when they are not specially excluded.   The principal genera and the best kinds are described in the following pages.

ALOYSIA.

The Sweet Verbena, *Aloysia citriodora*, is a plant with little show or decorative value ; but its sweet citron-scented leaves have given it a wide reputation, and its easy culture has established it as a favourite exhibition plant with the cottager and the amateur.   It is a free grower, which, with good management, will produce a fair specimen in two years from the cutting.   Pot in rich sandy loam and leaf-mould, set the plants in a frame near the glass, stake the leading shoot, and pinch the side shoots during summer to form a pyramid.   After the growth is made the water should be reduced, and the plants set in the open air to ripen the wood. Being deciduous, it should be kept almost dry to rest it in winter, and the shoots should be well pruned in previous to starting in a mild heat, earlier or later, according to the time it is wanted for exhibition. When fairly started, it should be grown in a frame or airy position in the greenhouse, and to have it in fine condition for the show, pinch and regulate the shoots to make a well-filled pyramid, and supply the roots with weak liquid manure to give breadth and substance to the leaves. The points of merit are : (1.) Size and symmetry of the specimen ; (2.) breadth, thickness, and health of the leaves.

ASPIDISTRA.

The *Aspidistra lurida variegata* is also eminently adapted for culture as an exhibition plant by the amateur, and, like the Sweet Verbena, it is often met with in fine condition in the cottagers' windows. It even takes a high place, when well-coloured, in any collection of greenhouse foliage plants, and for general usefulness in house decoration it has few equals. It is easily increased by division of the plant, and should be potted in a mixture of rough loam and peat, with some charcoal and bits of sandstone. Shift into larger pots as required without disturbing the roots, and give plenty of water in summer, with a little liquid manure to assist in developing the foliage to its full size, care being taken not to overdo it to spoil the variegation. Shade from strong sun, and give enough water in winter to keep the leaves and roots healthy.

The points of merit are : (1.) Size, health, and vigour of the plant ; (2.) size and colour of the leaves, which should be without blemish and the variegation pure.

AURICULA.

The Auricula, *Primula Auricula*, or Bear's Ear, better known in Scotland as the " Dusty Miller," has again become a very popular plant, after suffering neglect for a considerable period. Florists arrange them into Alpine and Stage Auriculas, and the latter are divided into Green-edged, Grey-edged, White-edged, and Selfs.

The following are some of the best varieties for competition in each class :—

Alpine.—A. F. Barron, Colonel Scott, Evening Star, Mrs. Dodwell, Sailor Prince, and Triumphant.

Green-edged.—Apollo, Colonel Taylor, F. D. Horner, Freedom, Lady Ann, and Prince of Greens.

Grey-edged.—Alderman, Charles E. Brown, Dr. Horner, George Lightbody, John Waterston, and Robert Traill.

White-edged.—Acme, Beauty, Conservative, Glory, Smiling Beauty, and True Briton.

Selfs.—Blackbird, Formosa, Heroine, Lord of Lorn, Othello, and Pizarro.

The points of a perfect flower of the Stage Auricula are, the tube yellow—not always obtained—circular, well filled with golden-coloured anthers, and free from fluting; the paste circular, solid, and as near white as possible; the ground colour unbroken and distinct where it surrounds the paste, but its outer edge flashes more or less into the edge, although the irregularities should never extend through it. The edge should be pure in colour, and the circumference of the pip should not be notched nor pointed, but perfectly even and smooth. Selfs have generally a notch in the lip of the petal, but some are without it, and the aim of all raisers is to have the "roseleaf" petal in all new varieties. Alpines should have a golden yellow or white centre, without powder, the body varied in colour, and the edge of one colour, paler at the outer edge.

Auriculas may be grown in the open border in ordinary soil, but to grow specimens for exhibition in their most perfect condition, a small house or frame, constructed expressly for them, is indispensable for success. In it they can receive abundance of light and air, two important agents in their successful cultivation, and be securely protected from storms of wind and rain. The plants should be regularly inspected, and their wants punctually met, so as to command the greatest success. The following compost suits them admirably :—Three parts of well-rotted turfy loam from an old pasture, one of dry cow-dung, one of leaf-mould, and one of sand, and in mixing the compost every worm that is seen in it should be picked out. Four-inch pots are generally sufficient, but the larger plants require 5-inch pots. In England it is usual to re-pot the plants in May, but in Scotland the best time is July and August. Use clean pots and drain them well. Fill in the soil and press it down at the sides of the pot, leaving a cone in the centre. Shake the earth clean from the roots of the plant, examine it carefully, and cut clean away all diseased or decaying parts, and trim the roots to about two inches in length. Dress the cut parts with powdered charcoal, and put some of it on the top of the cone, upon which place the plant, and spread the roots equally around. Press the soil firmly about the roots, and leave a quarter of an inch of the bare stem above the surface of the soil, which should be half an inch below the rim of the pot. To avoid bleeding, all leaves must be thoroughly withered before they are taken off. When re-potted, water the plants and put them back into the frame, and afterwards give water only when they require it, being very careful to let none drop into the heart of the plants. Remove them from the frame to the house about the end of September, where they should remain till they bloom, and when the trusses are well up they should be shaded from bright sun. Greenfly sometimes attacks the plants, but if the insects are at once washed off with a sponge they do little damage. In some places the plant is infested with a small grey louse called the " Woolly Aphis," which should be immediately cleared off whenever it appears, either at the collar of the plant or on the roots. Plants for competition should have a truss of five, seven, nine, or eleven pips, and before taking them to the show some cotton-wool should be placed between the foot-stalks to keep the pips from injury, and the truss tied to a small stake, these being removed when the show is reached. They should be securely packed in boxes to prevent injury in transit. In a collection of Auriculas, the plants should be even-sized, and the colours distinct. The points of merit are : (1.) Health of the foliage, height and strength of the flower stem ; (2.) the truss of flowers should be so disposed that the whole may be seen ; (3.) the pips should be round, flat, and free from blemish ; and (4.) they should be as near perfect as possible in colour and markings.

AZALEA.

The Ghent and Japanese (*A. sinensis*, syn. *A. mollis*) varieties supply diversity of colour and profusion of bloom at any period

during winter and spring, with very little trouble and at small
expense. A few of the best Japanese varieties are Apelles,
Baron E. de Rothschild, Consul Pecher, Flora, Phœbe, and
A. *sinensis alba grandiflora ;* and of Ghent varieties, Admiral
de Ruyter, Coccinea Major, Madame J. Baumann, Macrantha,
Princess of Orange, and Viscosa floribunda. They are easily
grown, fair success being attained by lifting healthy plants
well set with flower-buds from the open border, potting them
in sandy peat, and introducing them to the forcing-house.
Plants that have been well managed and established in pots
previous to forcing are, however, more reliable, and should
be preferred. In a brisk heat they will flower in about six
weeks, but too rapid forcing is detrimental to good quality.
Liquid manure should be given when the buds begin to swell,
and continued till the flowers are opening, when the plants
should be removed to the greenhouse and shaded. The points
of merit are : (1.) Size and health of the plant ; (2.) size and
form of flowers and profusion of bloom ; and (3.) substance of
petals, pureness of colour, and distinctness of marking.

BALSAM.

This makes a showy and effective exhibition specimen when
grown and flowered to perfection. Being an annual, it is raised
from seed, which should be selected from the best strain, sown
early in March, in shallow pans filled with fine loam, leaf-
mould, and a sprinkling of sand, and placed in a hotbed to
germinate quickly. When about two inches high the seedlings
should be potted into four-inch pots, set near the glass, and
kept shaded till the roots are running freely, when abundance
of air and light should be admitted to keep the plants dwarf
and compact. When the pots are well filled with roots, shift
the plants into two sizes larger, giving them richer soil. The
final shift into nine-inch pots should be given as soon as the
roots reach the side, using rich soil, and potting moderately
firm. When the flower-buds appear, the plants should be
placed in an airy greenhouse, fully exposed to the light, and
supplied with weak liquid manure twice a week; that made from
sheep-droppings and soot is excellent. The points of merit are :

(1.) Size, compactness, and freshness of the plant; (2.) quality and quantity of expanded flowers, which should be double, full, and well formed.

BEGONIA.

The species and varieties of Begonia are both numerous and useful. The tuberous-rooted section, to which our remarks apply, includes splendid varieties, which produce seedlings, if carefully selected, of equal merit, if not superior, to the parents. A year-old tuber of good size is a suitable subject on which to build an exhibition specimen. The roots should be shaken out in March, and potted in six-inch pots, using a compost of rich sandy loam, old manure rubbed down, some bone-meal, sand, and wood-ashes. Plunge the pots near the glass in a mild heat, and give the shoots all the light and air possible to ensure a sturdy short-jointed growth. Before the flower-buds appear, remove the plants to a warm greenhouse, re-pot, and train out the shoots, keeping the lower ones down over the edge of the pot. Pinch the strong shoots and cut out the weak ones, to admit light and air through the centre of the plant. When the roots are through the new soil, feed with weak liquid manure and shade from bright sun. The points of merit are : (1.) Size and healthy condition of the plant ; (2.) size and number of flowers ; and (3.) substance and colour of the flowers, which should be pure, bright, and distinct.

CALCEOLARIA.

The herbaceous or greenhouse kinds of Calceolaria have flowers so different from those of other plants, and the variety of colour is so diversified as to give them prominence among other subjects for exhibition. Seed from the finest strain should be sown in June or July, according to the time the plants are wanted in flower. When the seedlings are large enough, prick them in boxes four inches apart among sandy soil and leaf-mould, and when they have made six to eight leaves, shift into six-inch pots, using one part sandy loam, and one part leaf-mould, sand, rotted sheep or cow manure. Place near the glass in a cool frame on a damp bottom, water regularly, and keep them clear of insects. When the roots have reached the side of the pots, they should be shifted into nine-inch or flowering pots, giving no more heat than is necessary to keep out frost.   Early in spring give liquid manure regularly, made from sheep-droppings and soot, and an occasional pinch of good artificial manure spread over the pots once a week, and watered in with clean soft water, has a beneficial effect on the size and quality of the flowers.   Stake the plants neatly before the flowers begin to expand, keeping the lower side-shoots down to an angle of 45°, and the remainder equally distributed over the plant.   The points of merit are : (1.) Size and health of the plant, which should be dwarf and compact; (2.) number of heads and size of flowers ; and (3.) texture, colour, and form of the flowers.

CHRYSANTHEMUM.

An extensive and useful tribe of plants, which has representatives throughout Europe and Asia; the species from which the most popular flower of the day has sprung, *C. sinensis*, being a native of China.   The varieties have been crossed and re-crossed, with the result that all imaginable colours and forms have been produced, and it almost seems as if it were impossible to further improve or distort the race. The following dozen varieties are some of the best for culture as specimen plants for exhibition :—Alfred Salter, Belle Paule, Blanche Fleur, Bouquet Fait, Golden Christine, James Salter, Margot, Madame Bertie Rendatler, Madame de Sevin, Mrs. G. Rundle, Mrs. Dixon, and Soleil Levant.   These represent the various forms, and are all free-flowering, which is of the greatest importance in plants for exhibition.   The best time to take cuttings is from the middle of December to the middle of January, and the best place to strike them is in a cool house or frame where heat can be applied to keep out frost.   In selecting the cuttings, the sturdy, healthy, clean shoots at or near the base of the stem should be taken.   They should be inserted firmly in $2\frac{1}{2}$-inch pots, using a mixture of leaf-mould and river sand in about equal proportions.   They should then be set as near the glass as possible, and in a position where they can be shaded from the sun, and kept close until they are making roots freely, when they should receive all the light and air they can stand without flagging, so as to secure short-jointed sturdy growth. When the plants nearly fill the pots with roots, they should be shifted into pots an inch larger, using loam, leaf-mould, and sand, replaced in the same position, and carefully watered until the roots have penetrated the new soil. By the beginning of March all rooted cuttings should have received their first shift, and the most forward will be ready for the next, after which they should be placed in cold frames, protecting them only from frost and cold winds. It is a good plan to turn the plants carefully out of the pots and examine their roots before shifting them, as some varieties root quickly and may be ready for shifting by the beginning of March, while others may not require it until towards the end of April. Great care and discrimination is required at this stage of their growth, and any slow-rooting varieties should be brought forward with heat, to be finally shifted along with the others.

The second shift should be into 5½-inch pots, the strong growers into pots an inch larger, using soil similar to that employed at the last shift. When the roots have taken possession of the soil, the plants should receive their final shift into flowering pots, which may be 9 or 10 inches in diameter for single plants, and 11 or 12 inches when three plants are placed in a pot. Larger pots are unnecessary, and should not be allowed in competition. The compost should consist of two-thirds rich fibry loam, and one-third in equal parts of leaf-mould and dried cow-manure, with the addition of rough sand to make the whole free and open. To every barrow-load should be added one peck of bone-meal and one-half peck of any good artificial manure. In the case of extra-heavy soil, a little charcoal nuts should be added, and the whole well mixed together. A "potting-stick" should be used to firm the soil round the side of the pot, so as to produce hard fibry roots, which give firmness and solidity to the wood, a condition essential for the production of a quantity of flowers of good quality. Over-crowding in the frames should be avoided, and the sashes should be taken off in favourable weather, but replaced in frosty nights or during heavy rain.

If greenfly makes its appearance, the infested shoots should be dusted when dry with tobacco-powder; and if mildew appears, it should be immediately dusted with flowers of sulphur. The plants should be ready for their summer quarters by the 1st of June. An open airy position, but well sheltered from strong winds, is the best for them. They should be arranged at suitable distances apart, leaving sufficient room to get between them for disbudding, tying, and watering. There are various modes of training adopted for exhibition plants, such as pyramid, standard, and others, but unless a particular form is specially stipulated for in the schedule, the best, as producing good results with a minimum of labour, is the low bush form. The young shoots should be early started round the stakes in a spiral course, and pinched in the end of May or beginning of June, according to the earliness or lateness of the variety. Three shoots should be preserved from this break, which would give nine crown buds to the plant. This number being rather few, and possibly too early, these shoots should also be pinched after making about six inches of growth, and other three shoots preserved from each, thus giving twenty-seven terminal buds to the plant, which number is sufficient for a pot of large flowering chrysanthemums. All other shoots should be at once rubbed out, the branches tied well down to keep the plants dwarf, and the leaves all brought to the outside for exposure to the light. During July the growth will be rapid, and requires much attention. The plants should be looked over twice every day, giving water only to those actually requiring it; too much water does more harm to the plants than slight want of it, but they must not be allowed to become parched or to flag. Where the pots are exposed to the rays of the sun, they may be shaded with boards, or any suitable material at command. When the buds appear, they should be all nipped out except the terminal ones, which should be carefully preserved and tied into position, spreading them as equally as possible over the plants.

When the plants are thoroughly established and the pots full of roots, usually about the middle of July in the case of vigorous growers, they should be fed every second or third day with weak liquid manure, gradually increasing the strength of the liquid as they are able to bear it. We have adopted the following method and materials, with fair success, for feeding chrysanthemums. First, a barrowful of fresh cow-dung and a bushel of soot are put into a half-hogshead cask, mixing it well. The cask is then filled with water, well stirred, and allowed to stand to settle, when the clear liquid is fit for use, and is reduced to the required strength with clean soft water. When the clear liquid is exhausted, a barrowful of horse-droppings is added, and the cask re-filled with water, well stirred, and allowed to settle before using. In the second week in August the plants are top-dressed with about half an inch thick of the following mixture:—A third of turfy loam, another of leaf-mould and coarse sand, and the remaining third equal parts of bone-meal, artificial manure, and wood-ashes. Until the roots have fully occupied the top-dressing, only clean water is given; afterwards diluted liquid manure is again applied. About the first week of September the cask is cleaned out, and a barrowful of fresh sheep-droppings, the same of cow-dung, and a bushel of soot are put in, filling it up with water, stirring it well, and allowing it time to settle as before. It is then given to the plants at every watering, the strength being carefully regulated according to the condition and requirements of the plants. In about a fortnight afterwards a barrowful of manure from the poultry-yard, with a little sulphate of ammonia, are put into the cask, which carries on the supply till the beginning of October. The cask is then refilled with a mixture of cow, sheep, and pig's dung, adding some soot and artificial manure, which will last as long as liquid manure is beneficial. These have all proved excellent liquids, but they must be used with sound judgment, and inexperienced cultivators must remember that "weak and often" is a golden rule in the use of all stimulating manures, whether they are liquid or solid, and they should never be applied to plants suffering from dryness.

By the beginning of October the plants should be well

furnished with strong leathery leaves, covering all stems
and stakes, and the buds hard and well formed. As soon
afterwards as there is any danger from frost, the plants
should be taken into the greenhouse, given plenty of air in
favourable weather, and a little heat early in the day to
dispel damp. If necessary, the heat may be kept up gently,
to bring the blooms forward for the show, but the less of
artificial heat that is required the better for the blooms. The
points of merit are: (1.) Size and condition of the plant;
(2.) size, number, and freshness of the blooms; and (3.) colour
and substance of the petals.

CINERARIA.

The Cineraria is another very showy plant, which makes an
excellent companion to the Calceolaria, and succeeds under
similar conditions, except that it will not endure so low a
degree of cold with impunity. There are some excellent
named varieties, but plants raised from seed of a good strain
are all that can be desired. Seedling plants for exhibition
must be from a specially good selected strain, or they are just
so much labour lost, and end in bitter disappointment. Seed
should be sown about the same time as recommended for the
Calceolaria, and the plants grown on in the same manner.
The final shift should be into ten-inch pots, using a little
more manure and rougher soil than at previous shifts. They
are particularly subject to the attacks of greenfly, and should
be slightly fumigated at short intervals to keep them in check,
as they will not bear strong fumigation. The points of merit
are: (1.) Size and condition of the plant; (2.) size of flowers,
and number expanded; and (3.) breadth, substance, and form
of the petals, and distinctness of marking.

COCKSCOMB.

Although of a rather stiff and formal habit, this is a
favourite exhibition plant in many localities, and well repays
a little extra care to bring it to perfection. Being a stove
annual, the same soil and treatment that the Balsam requires
suits it well; but it must be grown throughout in stove-heat

to get it up to its best exhibition form. It should be kept close to the glass to keep it dwarf and stocky; but if it grows too high, that can be remedied by notching the stem just as the comb appears, on both sides near the base of the leaves, and " mossing " it with a handful of rich light soil enclosed within the moss, into which the stem will quickly root from the upper part of the notch. When the comb is fully grown, the stem will be well rooted in the moss, and then may be safely cut off below the ball, placed in a suitable pot, and shaded for a few days till the roots run freely into the new soil, when it should be gradually hardened off, and is then ready for exhibition. Combs of a very large size have been grown in an ordinary dung frame, when kept close to the glass and supplied with liquid manure twice a week. The chief points of merit are : (1.) Size of the comb and health of the foliage ; (2.) form of the comb, which should be close and even ; and (3.) substance and colour of the comb, which should be uniform.

CYCLAMEN.

The Persian Cyclamen forms a most beautiful and effective dwarf specimen when well grown, and it is specially attractive when used along with other plants as an edging to a group. There are many good named varieties, but plants can be grown from seed, the best of which will make first-rate specimens for exhibition the second season after being raised. The seed should be sown early in August, and the seedlings pricked off and grown on in rich light loam and sand with all the care that can be given to them, so as to secure well-ripened crowns two inches or more in diameter by the following autumn. They should be re-potted into five-inch pots, and started from September to November, according to the date of the show. They thrive well in a compost of rich fibry loam, leaf-mould, well-rotted manure, and silver sand, well mixed, and made only moderately firm in the pot. A little bottom-heat should be given until the roots are well through the soil, when the plants should receive a final shift into seven-inch pots, to which they should be confined in competitions. A slight shade from strong sun and a plentiful supply of water are essential to free growth. When they are nearly full grown and the flowers appearing, they should be set in an airy place near the glass in the greenhouse, and supplied freely with manure-water. The points of merit are : (1.) Size and perfect condition of the plant ; (2.) number and size of the flowers ; and (3.) quality of the flowers, which should be well formed, petals thick, pure, and rich in colour.

DEUTZIA.

The most useful of this genus are *D. crenata flore pleno* and *D. gracilis*. The latter is a most graceful plant, and the one generally chosen for exhibition. Beginning with a healthy specimen just done flowering, the shoots which have borne the flowers should be well cut back to encourage strong growths from the base. When they have grown about two inches, the plants should be turned out, the roots shortened, and re-potted in sandy loam, leaf-mould, well-rotted manure, and a sprinkling of bone-meal. The heat and moisture in a newly started vinery suits them well while making growth, and when it is finished the plants should be gradually hardened off, afterwards placed outside to ripen, and then stored in a cold frame for the winter. Before starting, the shoots should be regulated and trained into shape to form a free feathery bush, which is the best shape for exhibition, the shoots bending out gracefully, and so disposed that the flowers may be well seen. They should be started and treated as the Azalea for exhibition. The points of merit are : (1.) Size and health of the plant ; (2.) number and length of sprays ; and (3.) size and texture of the flowers.

DIELYTRA.

D. spectabilis is a showy hardy herbaceous perennial, well adapted for forcing for exhibition. Large stools should be planted three feet apart in a rich open border, and due attention paid to their proper treatment for two seasons, by carefully manuring, watering, and staking them, when they will be in the best condition for forcing. A 12-inch pot should be filled with one or more of the stools, using rich loam and well-rotted manure, with good drainage. After being watered, they may be started at once if the time is right for the show. The plants, when in full growth, should be liberally fed, and the flower-stems neatly supported by stakes. The points of merit are : (1.) Size and health of the plant ; (2.) number and length of the panicles of flowers ; and (3.) size of flowers and purity of colour.

GERANIUM.

Zonale Geraniums for exhibition purposes are usually divided into (1.) Zonales proper, or those with a more or less distinct dark zone on a green leaf ; (2.) Bicolors ; (3.) Bronzes ; and (4.) Tricolors, in accordance with the prominent features of their foliage. The Zonales are grown and exhibited for their flowers alone, and are generally subdivided into single and double-flowered varieties. The variegated-leaved, or Bicolor, Bronze, and Tricolor divisions, are exhibited solely for their foliage, and flowers rising above it are decidedly objectionable. A few excellent varieties of single Zonales are H. Cannell, John Gibbons, Meteor, Niphetos, Plutarch, and Vesuvius ; and of double varieties, Charming, Circe, Gambetta, Gloire de France, Mrs. H. Cannell, and Mrs. Langtry. Ivy-leaved geraniums may also be included here, because the great improvement in them in recent years fully entitles them to a special class at exhibitions. Some of the best varieties are A. F. Barron, Emile Lemoine, La France, Masterpiece, Madame Thibaut, and Charles Turner. Geraniums can be had in flower all the year round under good treatment, but the cultivation described here applies specially to those grown for exhibition at spring shows. The plants should be firmly potted in light loam, leaf-mould, sand, and rotted manure well rubbed down, with a sprinkling of bone-meal and wood-ashes. The shoots should be tied down to wires running outside the pot, two or three in number, according to the size of the plant. A light airy position in the greenhouse should be given them, and the strong shoots stopped till the specimen is complete ; when the foliage should round off nicely to about three inches below the rim of the pot. Liquid manure should be given when growth is being made, and the flower-trusses picked off as soon as they appear, until three or four weeks, according to the time of the year, before the plants are wanted for exhibition. The points of merit are : (1.) Size, form, and health of the plant ; (2.) number and size of the trusses, which should stand well above the foliage ; and (3.) quality and size of the flowers, which should be fresh and pure in colour ; the single large and broad in petal, and of good texture ; the double well formed, full, and each pip standing out clear, without crushing in the truss.

Some of the best Bicolors are Chelsea Gem, Elegance, Flower of the Day, Iduna, Mrs. Parker, and Triomphe de Gand. Of Bronzes: Best Bronze, Black Douglas, and Maréchal M'Mahon. Of Golden-leaved : Cloth of Gold, Crystal Palace Gem, and Golden Superb. Of Golden Tricolors : Lucy Grieve, Masterpiece, Mrs. Henry Cox, Mrs. Strang, Prince of Wales, and Queen of Tricolors. And of Silver Tricolors : Lass o' Gowrie, Her Majesty, Isle of Beauty, Italia Unita, Mrs. Laing, and Queen of Hearts. Their cultivation is similar to that of the Zonale section, except that the soil should not be so rich, and being more difficult to keep healthy during the winter, they should be placed on a shelf near the glass in the warmest end of the house, and no more water given than will keep the roots from shrivelling. The Bicolors and Tricolors should be grown in a cold frame during summer, and kept close to the glass. The Bronze and Golden-leaved may be set on boards in the open air in a sheltered place. When the pots are well filled with roots, liquid manure may be given to increase the breadth of the foliage, but not sufficient to affect their perfect colouring. All flowers should be removed as they appear, and the shoots tied in and pinched as they require it. The points of merit are : (1.) Size, health, and form of the plant ; (2.) colour and freshness of the leaves ; and (3.) colour of the zone, which should be distinct and bright, and the light parts of the leaf pure and clean. The points of merit in the Bronze section are similar, except that the zone should be a light or dark bronze, and distinct from the other parts of the leaf, which should also be of a bronze shade ; while the Golden and other kinds should conform exactly to their names.

GLOXINIA.

There are many good varieties of this popular flower in cultivation, from which the best for exhibition can easily be selected, those with erect flowers being generally most appreciated. Great variety of colour and excellent form can be obtained from seed, which should be sown in pans of fine peat and leaf-mould during February, and placed in a temperature of 65°. Care should be taken not to cover the seed, which should be sown on a thin layer of pure silver sand, pressed down, and watered through a fine rose. Cover the pan with glass and shade the young seedlings from sun, and as soon as they can be handled prick out in boxes, and admit sun and air after they have started to grow. Their subsequent treatment is similar to that of the Begonia, except that fibrous peat should be substituted for loam, and they require a higher temperature to develop the foliage and flowers. The best varieties among the seedlings should be selected in autumn, and well ripened preparatory to being cultivated for exhibition specimens the following season. The points of merit are: (1.) Size of plant, health of foliage, and number of flowers; (2.) size and substance of flowers; and (3.) colour, which should be pure and bright, and the marking distinct.

GUELDER ROSE.

This well-known and conspicuous hardy flowering shrub is one of the most useful for early spring forcing, and when a good specimen is seen at a flower-show it attracts much attention, with its profusion of globular heads of snow-white flowers. Suitable plants should be well prepared beforehand, by a year or more of good cultivation in rich loam, in a warm sheltered spot, fully exposed to the sun, where the plant can be carefully built up into an exhibition specimen by attention to watering, manuring, and regulating the shoots, which should be thoroughly ripened the autumn previous to the show. The plants should be lifted and potted in good sound loam as soon as the leaves have fallen in October, and then set away in a cool place, protected from storms until the time arrives for

starting them.   About eight weeks careful forcing in spring
is sufficient to bring them into flower.   Great care is required
to regulate the opening of the flowers, so that they may
carry safely to the show, which must be within a few days
after they are fully expanded.   The points of merit are: (1.)
Size and vigour of the plant; (2.) size, freshness, and number
of heads of flowers; and (3.) colour pure white, and petals of
good substance.

HYACINTH.

There is a great choice of good varieties among these, and
every competitor has his own particular fancy.   A dozen of
fine varieties for exhibition, including those with blue, red,
white, and yellow flowers, are Alba maxima, Czar Peter,
Grandeur-à-Merveille, Ida, King of Blacks, King of Blues,
King of Yellows, Kohinoor, La Grandeur, Lord Macaulay,
Solfaterre, and Von Schiller.   Only well-ripened firm bulbs
can produce first-rate exhibition flowers.   They should be
potted in October to be in flower in April.   Use five-inch pots,
well drained and filled with a rich compost of light loam, leaf-
mould, sand, and well-rotted manure, placing the bulb about
two-thirds into the soil.   Water, and then plunge the pots in
ashes in the open air, covering them six inches deep.   Remove
them to the greenhouse when about two inches started, shading
slightly till the leaves become green, and then give air freely
to keep them sturdy.   As soon as the flower-scapes appear
the plants should be regularly fed with weak liquid manure,
made from sheep-droppings and soot, and given an occasional
sprinkling of guano, or any good artificial manure, on the
surface of the soil, and watering it in.

The spikes of flowers should be attended to as they grow,
and all pips put into place, so that the head may be even and
regular.   The flower-stem should be so stout as to keep the
spike erect without the aid of a stake, but for safety in transit
to the show it is necessary to support each flower-spike with
a neat stake.   The spikes should be carefully dressed, and
all pips placed in proper position by the use of two pointed
sticks before they are staged.   The points of merit are:

(1.) The length, diameter, and symmetry of the spike; (2.) the size and substance of the pips, which should be well shaped; and (3.) the purity, and richness of the colour.

HYDRANGEA.

A genus of hardy or greenhouse plants, some of which are largely cultivated for their handsome heads of bloom. They are naturally autumn-flowering, but with special treatment can be brought into flower early in the season. The best for exhibition are *H. hortensis* and *H. paniculata grandiflora*. The former produces large heads of rose-coloured flowers, while the latter has long spikes of beautiful pure white flowers. The proper ripening of the wood and early setting of the flower-buds are the two points which must be secured to ensure successful forcing and early flowering. As soon as the flowering is past, the plants should be cut down to two eyes from the base of the shoots, re-potted, placed in heat to make their growth, and afterwards transferred to an airy position near the glass in the greenhouse till the wood is ripe and the buds prominent. They should not be kept too dry during winter, and may be forced into flower under the same conditions as the Deutzia. The points of merit are: (1.) Size and health of the specimen; (2.) number and size of heads of flower; and and (3.) size, substance, and colour of the flowers.

LILAC.

This is one of the most beautiful and useful of hardy flowering shrubs, and forces well for exhibition in early spring, when the plants are properly prepared, the best kinds being Charles X. and the Persian. The first requires a place in the forcing-house very much shaded, in order to bring out the pure white colour which it assumes when forced, otherwise they should be treated the same as the Azalea, except that equal parts of loam and peat should be used in potting. The points of merit are: (1.) Size and condition of the plant; (2.) size of spike, and of flower; and (3.) richness, and purity of colour.

LILY OF THE VALLEY.

This chaste flower is a general favourite on account of its modest beauty and sweet perfume. To obtain strong crowns for forcing, as near as possible equal to those grown in Germany, the plants should be cultivated on rich firm ground, in a sheltered position not too much exposed, planting the crowns singly two inches apart and in rows one foot asunder. It is of the greatest importance that the leaves have plenty of room to develop and the crowns to get well ripened. In selecting the crowns for an exhibition specimen, the rows should be loosened with a fork, and the strongest drawn out with roots attached. They should then be thickly set in a shallow pan 14 inches wide, using a compost of rich loam, leaf-mould, and sharp sand. They should be watered and placed in a close frame for two or three weeks previous to forcing. They like a strong moist heat, and plenty of light to develop the leaves along with the flowers. The points of merit are: (1.) Size and health of the specimen; (2.) number and length of the flower-spikes; and (3.) size, purity, and freshness of the flowers.

MIGNONETTE.

This plant is grown chiefly for the fragrance of its flowers, and is esteemed by many as of more value than a gorgeous display of brilliant inflorescence. To obtain large exhibition specimens, seed from a vigorous and choice strain should be sown about the beginning of March in three-inch pots, among sandy soil and leaf-mould, and placed in a gentle heat. When the seedlings are an inch high, all but one should be pulled out, leaving the best, which should be staked and shifted into a larger pot as it increases in size, using as compost three parts sandy loam and one part rotted manure, with a dash of lime and soot, which sweetens the soil and improves the colour of the leaves. The plants may be trained as pyramids by pinching and staking, or in the umbrella form, which is the most artistic, and is performed as follows:—The plants should be placed in the greenhouse and grown with a single stem to the height of two feet, preserving the leaves, but removing all side-shoots as they appear, except three at the top, which form the foundation of the specimen. A stout stake should be placed beside the stem, and two strong wires, each $2\frac{1}{2}$ feet long, stretched at right angles across the top, with the ends slightly bent downwards, to which is attached a circular wire, which forms the outline. Thin wires should then be tied over the cross ones, to form a trellis. The shoots should be stopped, tied down, and all flowers removed until within a month of the time the plants are wanted for exhibition. The points of merit are : (1.) Size and health of the specimen; (2.) length and number of the spikes ; (3.) clear colour, and sweet perfume of flowers.

NARCISSUS.

These popular spring-flowering bulbous plants are divided into two sections, the Garden and the Polyanthus Narcissus. The former is distinguished by having one, two, or three flowers on the spike, while the latter has many, but geneally of a small size. A few of the best Garden varieties are Empress, Emperor. Horsfieldii, Maximus, Princeps, and Sir Watkin. Some of the best Polyanthus Narcissus are Bazelman Major, Grand Monarch, Her Majesty, Milton, Newton, and Queen of the Netherlands. Their cultivation is somewhat similar to that recommended for the Hyacinth. The bulbs for exhibition plants should be carefully selected, by choosing all of the same size, firmness, and appearance, because it is of the greatest importance to have all the flowers in a pot fully developed at one time. From three to six bulbs should be put into a pot according to its size, among light turfy loam, sand, and leaf-mould. The pots should be covered with ashes like the Hyacinth, till they are partially filled with roots, when they are in a proper state for forcing. The plants are much benefited by weak liquid manure when they are fairly started. The points of merit are : (1.) Size and texture of the petals ; (2.) prominence of the trumpet, the edges of which should be slightly reflexed; (3.) the colour clear and rich. If the perianth and trumpet be of different colours, they should be distinct. In Polyanthus Narcissus, all other points being equal, the number of flowers on the stem is the chief point to decide their merits.

H

PELARGONIUM.

When well grown, these form some of the showiest of spring and early summer exhibition specimens. The varieties are extremely numerous, but the following dozen includes some of the best show, fancy, and regal sections:—Achievement, Beauty of Oxton, Brilliant, Captain Raikes, Dr. Masters, Exquisite, Mrs. Harrison, Mrs. Mathers, Penelope, Queen Victoria, Triumph, and Volonté Nationale alba. They grow well in three parts of fibry loam to one of leaf-mould, with a sprinkling of decayed manure, bone-meal, wood-ashes, and sand, using a little coarse silver sand for the weak-growing varieties. An airy place near the glass in the greenhouse is the best for them, and the growth should be tied out and pinched, so as to furnish the plant all over with shoots of equal length and strength. They should be regularly fed with various kinds of liquid manure previous to flowering, and fumigated often to keep down greenfly. After flowering, the plants should be placed in the open air, and kept moderately dry, to ripen them. They should then be cut close down, watered, and placed in a close frame, to start them about half an inch, when they should be re-potted, and air given freely to promote sturdy growth, to flower the following spring. The points of merit are : (1.) Size, form, and condition of the plant ; (2.) size, quality, and quantity of the flowers.

PETUNIA.

This is a very showy plant when well cultivated, and is admirably suited, from its character and habit of growth, for growing into a shapely exhibition specimen. Some of the best double and single varieties are Arlequin, Derviche, Louis Ratisbonne, Marc Alban, Panama, and Phœbe. To obtain fine plants with large flowers, they must be young, as old plants very soon lose their freshness and size of flower. They grow very rapidly, and cuttings early rooted may be grown into good plants the same season, but summer cuttings grown into stubby plants before winter make the best plants for autumn shows. After the cuttings are well rooted, they should be potted in light rich soil with leaf-mould and sharp

sand. They should be set near the glass, and kept moving
during winter. In early spring the plants should be re-
potted, and afterwards shifted on as they require it, using
richer soil, and feeding the roots as the soil becomes exhausted.
The shoots should be pinched until sufficient are obtained to
make a full specimen. Instead of staking, they should be
trained over wires fixed outside the pot, in the same manner
as the Pelargonium. Near the glass, in a cold frame, is the
best place to grow them in summer, and if they are regularly
pinched and watered with liquid manure twice a week, excellent specimens should be the result. The shoots should be
tied in, and all flowers pinched off till within three weeks of
the time the plants are wanted in flower. The points of
merit are: (1.) Size, form, and health of the plant; (2.) size
and number of flowers; and (3.) quality, and colour; the
petals being thick and of good texture, and the colours bright
or pure.

POLYANTHUS.

A pretty and interesting class of florists' flowers, originating
probably from a cross between the Cowslip and the Primrose.
To grow them well in pots for exhibition, they require the
protection of a frame, and to be treated otherwise much in
the same way as the Auricula. They should be potted in
August, shaking the soil carefully away from the roots, and
cutting off all signs of rot or decay, to which they are often
liable. Any that do not need re-potting should be top-dressed
in February with good rich loam, leaf-mould, and dry well-
rotted manure. When newly potted, keep the frame close for
a few days until the roots run through the soil and the plants
begin to grow, when the sashes may be gradually removed
and the plants be fully exposed to the light and air. The
frame should be constructed to allow a free current of air to
play over the plants even when the sashes are on, except
during severe frost. Water should be carefully given at all
times, but particularly in winter, sufficient being given to
saturate the soil, and no more, until it is really required,
taking care that the plants are not allowed to flag, and that

no water falls into the hearts of them to induce disease. The plants may be divided when shifted in August to increase the stock, but excellent varieties are easily raised from seed, which are often as fine in quality as named ones, and are much less liable to go off suddenly with rot, so that they are preferable.

The properties of a show Polyanthus resemble very much those of the Auricula, with the difference in the colour and markings. The tube should be short and well filled with anthers ; the eye should be round, of a bright clear yellow, and distinct from the ground, which should be shaded light to dark crimson with a stripe in the centre of each limb from the edge to the eye, where it should end in a fine point. The edging should be gold-lace, clear and distinct, and of the same colour as the eye and strips. The pips should be large, flat, and round.

PRIMULA.

The Chinese Primrose, *P. sinensis*, and *P. obconica* are two of the most useful of spring flowering plants. The shades of colour and size of flowers in the varieties now introduced are exceedingly pretty, and when exhibited *en masse* have a charming effect. For flowering in spring, the seed should be sown in the previous June in pans of leaf-mould and sand, covering with the latter, and plunging in a temperature of 55° to 60°. When the seedlings appear, set the pans near the glass and shade from strong sun, and when large enough, prick them off into small pots, using the same compost. When the plants are a fair size and the pots full of roots, they should be finally shifted into six-inch pots for flowering, using as compost one half sandy loam, the remainder leaf-mould, dry sheep or cow manure, and sharp sand, well mixed and used in a fine state. Tie the plants to short stakes to keep them steady, and after they start to grow a cool frame suits them best till September, when they require to be placed in a greenhouse. Keep the plants moderately dry and the flowers pinched out during winter. In early spring they should be encouraged to grow vigorously with weak liquid manure, increasing its strength as the time for flowering approaches. The flowers should be allowed to expand about a month before the show, and no effort spared to have the plants in full bloom by the end of that time. *Primula obconica* is a continuous bloomer, and is much easier had in perfection under similar treatment than its companion. The points of merit are : (1.) The size and condition of the plant ; (2.) size and symmetry of the heads of bloom ; and (3.) the flowers fully expanded, and the petals of good colour, form, and substance.

RHODODENDRON.

This is an extensive genus of useful showy flowering shrubs, which, according to their different characteristics, form distinct groups. There are many beautiful shades of colour among their flowers, and most of them can be freely forced. The following are among the best hardy kinds suitable for exhibition specimens :—Austin Layard, Catawbiense, and its varieties, Frederick Waterer, Gloire de Gandavensis, Lord John Russell, Maculatum, Nobleanum, Præcox and its varieties, Robert Marnock, Sir William Armstrong, William Austin, and William Ewart Gladstone. Hardy Rhododendrons may be cultivated in pots specially for forcing, or they may be planted on a border made up of sandy peat, leaf-mould, and well-rotted manure. They are more easily kept in a good healthy condition in this manner than when confined in pots. They can be lifted for forcing with good balls, and potted without receiving a check in the operation. They should be potted in the same soil as that in which they are grown, and forced in the same manner as recommended for Azalea. The points of merit are: (1.) Size and health of the plant ; (2.) freshness and quantity of flower ; and (3.) colour, size, and substance of petals.

RICHARDIA.

The popular Lily of the Nile is a valuable decorative plant, and also makes a grand specimen for exhibition. To produce a large specimen quickly, a number of roots should be put together in a twelve-inch pot among rough sandy loam, peat, and pieces of charcoal. The plants should be placed in the greenhouse, supplied with plenty of water, and well fed with liquid manure. They bear forcing well, and can be had in full flower at any date in spring and early summer. The points of merit are : (1.) Size and health of the specimen ; (2.) size, number, and purity of the flowers.

ROSE.

The Rose, the Queen of Flowers, is always appreciated, but more especially after the dull winter months, and its culture in pots for exhibition specimens should be more practised than it is. All the different sections may be forced with fair success; but the Teas, with all their beauty and fragrance, are most highly valued, and most amenable to the influence of artificial heat. Some of the best tea-roses are Beauté de la Europe, Catherine Mermet, Duchess of Edinburgh, Etoile de Lyon, Madame Bravy, Madame Jules Margottin, Madame Sertot, Marie Guillot, Marie Van Houtte, Perle des Jardins, Reine Marie Henriette, and Souvenir d'un Ami. And of Hybrid Perpetual Roses: Abel Carrière, A. K. Williams, Alfred Colomb, Baroness Rothschild, Camille Bernardin, Charles Lefebvre, Dean of Windsor, Duke of Teck, Grandeur Lyonnaise, Marie Baumann, Pride of Waltham, and Souvenir de Madame Berthier. To produce large specimens for exhibition, a preliminary course of culture extending over several years is necessary, which, in many cases, is too long to wait, and therefore large healthy plants should be lifted in spring, pruned, and potted in twelve-inch pots among rich marly loam, sand, and rotted manure. They should be grown in a cool house, and be carefully watered and attended to otherwise, thinning the growth if crowded, and keeping the shoots free from insects and mildew. The weak-growing varieties should be on the brier or Manetti stock, while the strong-growers do well on their own roots. Previous to starting, tea-roses should be pruned in moderately, and the pots top-dressed with marly loam, well-rotted manure, and bone-meal. A night temperature of 50° is sufficient to start with, but should be increased to 60°, with a corresponding rise in day temperature, as the plants advance in growth. Plenty of air should be given in favourable weather, and weak liquid manure supplied until the buds begin to open.   The plants should be neatly staked out in bush form, keeping the stakes well hidden, the foliage free, and the flowers standing boldly out.   Hybrid Perpetual Roses require the same treatment as the Teas, except that they should be pruned closer, and the shoots tied out and kept in position without stakes.   The points of merit are : (1.) Size and health of the plant; (2.) number and quality of the flowers; and (3.) richness and purity of colour and fragrance of blooms.

SPIRÆA.

This is a hardy and extremely useful tribe of plants for decorative purposes, of which *S. japonica* and *S. palmata* make beautiful specimens for exhibition, the feathery flowers of the first being pure white, and of the second crimson.   They are easily forced, especially the former, and can be had in flower for several months.   Strong well-prepared specimens may be lifted from the open ground and potted in rich loam, leaf-mould, and sand, and placed in a forcing-pit.   When the spikes appear, the plants should be removed to a greenhouse and supplied with weak liquid manure.   When done flowering, the plants may be broken up, and should be planted in a rich open border, where they should remain for two years to prepare them for being again lifted and forced for exhibition. The points of merit are : (1.) Size and condition of the plant; (2.) size. number, and freshness of the flower-spikes.

TULIP.

These bulbous plants are divided into several sections, viz., Early, Late, Double, and Single Tulips.   The following are some of the best for pot culture and exhibition :—Double Flowered : Blanch Hative, Clothilde, Kaiser Wilhelm, Premier Gladstone, Rosa Bonheur, Violet Superior.   Single Flowers : Catafalque, Cerise Blanche, Comte de Nassau, Grand Monarch, La Superb, and Marie Theresa.   They succeed well under the same treatment as the Hyacinth, and should be freely fed with liquid manure when the pots are filled with roots, to bring the flowers to the highest state of perfection; but when the flowers begin to open, clear water only should be given. They should be neatly staked and slightly shaded, to preserve the colour and freshness of the blooms. The points of merit are : (1.) Size, equality, and form of the flowers ; (2.) arrangement and brightness of the colours, which should not run into each other, the petals being similarly marked, or if of one colour, it should be pure and constant ; and (3.) double flowers should be large and full, the petals regular, and of good size and substance.

GROUPS OF PLANTS.

These form an important feature in a flower-show, and if the plants have been selected with discrimination and are tastefully arranged, they never fail to command the attention and favourable comment of the public. In considering the selection of the best plants and their arrangement to the best effect in a group on a stand or table 20 feet long by 7 feet wide, a commencement should be made with three leading plants of graceful form, standing well over the main body of the group, and arranged along the centre, say a well-coloured Croton of elegant habit in the middle, and about five feet off, towards each end, a graceful specimen of *Cocos Weddelliana* or other palm ; or a palm, say *Phœnix rupicola*, may be the middle plant, supported by two Crotons or two fine specimens of Dracænas towards the ends. A groundwork should next be formed of dwarf ferns—*Adiantum cuneatum* in small pots is among the best—arranged to cover the stand and effectually conceal the pots of the other plants. The prominent plants should then be placed in position with all the judgment and taste possible, regulating their height above the groundwork in such a manner as to display the specimens to the best advantage. A choice selection of Alocasias, Anthuriums, Crotons, Dracænas, Marantas, Palms, Pitcher Plants, and such like, with ornamental and highly coloured foliage, should be freely interspersed with the choicest specimens in flower of Amaryllis, Azaleas, Eucharis, Geraniums, Fuchsias, Liliums, Orchids, and the like. After the main features have been arranged in a light and elegant manner, the edge of the table should be treated in the same way, using both flowering and foliage plants of a dwarfer habit, and finishing off with nice plants of *Ficus repens*, Fittonia, variegated Panicum, and Selaginellas, interspersed with drooping Fuchsias, Lobelias, *Thunbergia alata*, and the like, all set at regular distances, and drooping tastefully over the edge. The table should be well balanced without being too formal. The prominent plants should be in perfect condition, and so thinly set that the eye can clearly trace their outlines. Crowding should be strictly avoided, as a free open airy arrangement is in better taste, and most highly appreciated. The chief conditions to be taken into consideration are the cultural merits and high quality of the specimens, and the harmonious combination of the whole group.

The judging of tables for "effective arrangement" is not easily described in a satisfactory manner, because it is chiefly a matter of taste, which varies with the individual. Some appreciate bold contrasts, others prefer mellow tones; but what is graceful and handsome is approved by all. A miniature of a tropical forest with the feathery palm and slender grasses shooting out from an undulating surface of moss and fern, is a good basis to work upon, with clumps of bright-coloured foliage and flowering plants tastefully scattered about, and orchids peeping from among the branches or suspended in as natural a way as possible on the taller plants, making a very pleasing and attractive scene, and, when done with skill and good taste, goes a long way to secure high honours. Whatever idea or design is employed for such a display of taste, the plants used should always be appropriate, and perfect in leaf and flower, each occupying its proper space, without the formality of one exactly the same being placed in a corresponding position. The arrangement should be rounded off and subdued in such a manner that the eye may not be particularly attracted to any one part, but that it may wander with delight and satisfaction over the whole.

GENERAL REMARKS.

In concluding our remarks on exhibition plants, a few useful hints may be given on the cleaning, training, packing, and other points, of which the beginner often feels the need of information upon. The exhibitor must always remember that flowering plants are in their very best condition for competition when the first flush of flowers is fully expanded, and every effort should be made to secure that desirable point. They can be retarded by keeping them cool and shaded, as much as can be safely done without injury to the flowers; or when they are late, they may be forwarded by a judicious amount of extra heat.

Geraniums, Heaths, and all kinds of plants, and especially climbers, that require to be trained into shape, should be manipulated in sufficient time before the show to allow the leaves and points of the shoots to assume a natural position. Preparatory to exhibition, all foliage that will bear it should be carefully sponged over with a weak solution of soft soap and tepid water, to remove every speck of dirt; at the same time cutting out all discoloured and decaying leaves. Staking and tying should be done as neatly as possible, and no more of either employed than is actually necessary.

Foliage plants with single stems, such as Dracænas, Palms, and Cycads, and also Tree-Ferns, should have all their leaves laid evenly up to the centre, covered with tissue-paper, and carefully tied to a strong stake, with wads of cotton-wool inserted to prevent rubbing, to prepare them for safe transit to the exhibition. Tender flowering plants, such as Orchids, and stove plants generally, should have the flower-stem covered with cotton-wool and tissue-paper, and the separate flowers properly packed in the same material, finishing off by enveloping the whole truss in paper neatly tied with soft matting. Ferns should have the outside fronds laid up to the centre ones, and a breadth of soft cloth run round the plants, bringing it over the top, forming a protection to the tender fronds. When packing to travel by road or rail, soft meadow-hay is the best material to put round and below the pots, being careful to economise space by packing the smaller amongst the larger plants, placing them so that the leaves will not rub against anything.

Some exhibitors think their object is accomplished when the prize is won, and do not bestow much care in the removal of the plants from the exhibition. This careless method must be emphatically condemned, because any defect in packing and conveyance tells on the plants and on future success, and as much care and trouble should be bestowed on the removal of plants from an exhibition as will ensure their safe return to their home.

Where there is proper accommodation for growing exhibition plants, every specimen should have room to fully develop. It is a good practice to group the various kinds according to the treatment that suits them best. Crotons should be grown where they can get most light, with provision for slight shading during extreme heat. Ferns, again, should be kept in the shade and never syringed overhead, but moisture should be kept constantly amongst them by evaporation from damping the stages and paths. Filmy ferns, however, should be regularly "dewed" overhead with the syringe. All plants should be examined at least once every day to see if water is required at the roots. This operation is best performed in the afternoon in summer, and in the morning in winter. Careful attention should be paid to giving proper ventilation at all seasons. The admission of air by the bottom ventilators alone to all the hottest houses prevents the too rapid escape of moisture, which would leave the atmosphere parched and dry and quite unsuitable for growing many exotic plants. The ventilators should be large enough to admit the necessary amount of air, combined with shading, to keep the temperature within the prescribed limits. During the time of cold winds in spring, ventilation should be confined to the sunny side of the house, closing early in the afternoon to economise heat, and at the same time thoroughly damping paths and stages.

It is a good plan to syringe such plants as Crotons with a solution of two ounces of soft soap to the gallon of water, applying it as hot as it can be used with comfort. This practice serves to keep insect pests in check ; but when plants are badly infested, a wine-glassful of paraffin should be added to each gallon while the liquid is boiling, and allow it to cool down to a safe temperature before proceeding to use it. Another method is to add the paraffin to the soap and water, and then syringing alternately into the pail and on to the plant requiring to be cleaned. The higher the temperature at which the soap, water, and paraffin are mixed the better they amalgamate; hence they should be mixed at a temperature as near as possible to the boiling-point.

A solution of one part of soft soap to eight parts of water "takes" paraffin freely at or near the boiling-point, and forms the basis of the best insecticide. When properly used, paraffin is the best material for cleaning plants from all dirt and insect pests at an infinitesimal cost. It should, however, be always used with caution, and never applied recklessly or indiscriminately to all plants alike. It should not be used on delicate ferns or on tender young growths, as it is liable to injure them even with the most careful manipulation; but on such smooth-leaved and robust plants as Crotons, Dracænas, Gardenias, Palms, Stephanotis, and such like, which are so liable to the attacks of mealy-bug and scale, it may be applied with perfect freedom and with the best results. Repeated applications, at a strength of one gill of paraffin in the gallon of liquid, will eradicate the most inveterate of plant insect pests. In applying it to Azaleas and all downy-leaved plants, a very weak solution must be used, about a quarter of a gill of paraffin to the gallon, repeating the application till the plants are clean.

Most kinds of plants are best as well as more easily moved to and from an exhibition, when firmly potted in pots of a comparatively small size. A proper application of artificial and liquid manures will grow them to perfection in such pots, in which, as a rule, plants flower more profusely, and the foliage assumes a richer and deeper colour and more beautiful tints.

Soil, and all the materials for potting, should always be got ready some time before they are required for use, and kept stored in separate bunks in the potting-house, so that they may be at hand when wanted, and in the best mellow and warm condition.

DIVISION II.

CUT FLOWERS.

This is an important part of all horticultural exhibitions, and forms a centre of great attraction to visitors generally, while the florists' flowers are a source of unmeasured enthusiasm on the part of the amateur, who enters with the keenest zest into competition with his rivals in this interesting section. Immense improvements have been developed in recent times among the beautiful subjects comprised in this division, particularly in some of the florists' flowers classes, in which there has been a marked advance in the size and shape of the flowers, substance of petal, richness and variety of colour, and all other points that go to constitute a perfect bloom. It is not necessary, however, to dwell at length on the advances made by the various sections of cut flowers, and only a selection of the best varieties of each genus is given under their respective popular titles in the following descriptions.

ANNUALS.

Collections of the flowers of annuals have become very popular at autumn flower-shows, and when they are tastefully set up in proper-sized bunches they make an effective display. They are divided into two classes, half-hardy and hardy annuals. Some of the most showy of both classes, however, are useless for exhibition, as they fade within a short time after they are gathered.

HALF-HARDY ANNUALS.

These include the following beautiful species and varieties suitable for exhibition :—*Alonsoa Warscewiczii, Calliopsis*

chinensis, Gaillardia picta Lorenziana, Tagetes erecta; German
Asters, varieties; Marigold, French and African; Petunia,
varieties; Phlox Drummondi, varieties; Rhodanthe, varieties;
Salpiglossis, varieties; Matthiola (stocks), varieties; Tropæo-
lum, varieties; and Zinnia, varieties. The seed should be
sown in February or March in pans of light sandy soil and
leaf-mould, and placed in a temperature of 55° to 60°. The
seedlings should be pricked out as soon as they can be handled,
and afterwards be carefully hardened off for planting out in
May and June.

HARDY ANNUALS.

Many species of these are so rich in varieties of merit that
it is scarcely possible to name a selection of them that would
prove to be the finest in every season. The following dozen
species, however, with their varieties, form a collection to
be depended upon :—*Calendula officinalis, Calliopsis tinctoria,
Chrysanthemum tricolor, Clarkia pulchella, Eschscholtzia cali-
fornica, Godetia elegans, Iberis umbellatus, Lathyrus odoratus,
Lupinus nanus*, and other species; *Nigella damascena; Papaver*,
several species and varieties.

The culture of hardy annuals under ordinary circumstances
is a simple matter, as they grow freely in any rich soil of an
open texture. They may be sown any time from the middle
of February till the end of May, according to the state of the
weather and the date at which it is desired to have them
in perfection for exhibition. From twelve to fourteen weeks
should be allowed in ordinary seasons between the time of
sowing and the date of the show. As a rule, it is best to
make two or three separate sowings, at intervals of eight
days, so as to have them in perfection at a given date. Small
seeds (Mignonette) should be covered to a depth of about
twice their own diameter, larger seeds (Candytuft) three or
four times their diameter, and for large seeds (Sweet Peas)
from one to two inches is the proper depth. As soon as
the young plants are large enough to handle, they must be
well thinned out, to admit light and air among them, which
promotes strong sturdy growth and well-developed flowers.

If the flowers are likely to come too early, a slight shading will retard them, but it must be cautiously done, so as not to draw the growth or spoil the colour of the flowers. In dry weather copious waterings are required to keep the plants fresh and vigorous, and a dose of weak liquid manure occasionally is beneficial. For competition, the flowers should be cut early in the morning, and with as long stalks as possible, so that they may last the longer, and be set up effectively. The bunches should be made up as light and free as possible, to show every spike and flower clearly. In most cases a single colour (one variety) will be found most effective; while in others several colours (varieties of one species), when properly combined, will produce the most charming effect. The points of merit in annuals are : (1.) Freshness and tasteful arrangement of the bunches; (2.) size and texture of the blooms; and (3.) brilliancy and purity of the colours.

ANTIRRHINUM.

This forms a prominent object at many local flower-shows in the autumn months, and is a special favourite in the cottage-garden. Among the best are Anson Shiel, Agnes, Cleopatra, Favourite, Henry Wood, and The Bard. To obtain exhibition specimens the plants should be raised from cuttings annually. Select stubby, short-jointed shoots early in the autumn, and insert them in sand and leaf-mould, placing them in a cold frame near the glass, and in such a position as not to require shading. By April they will be well rooted, and should then be planted out two feet apart on ground that has been well pulverised, manured, and specially prepared for them. Keep clean from weeds, and reduce the number of shoots if too many are produced. Stake them when necessary. Mulch and water in hot dry weather. The points of merit are : (1.) Number of flowers and size of spike; (2.) size and texture of the tube, the lip of which should be well extended to show the throat; and (3.) the colour clear, the markings well defined, and the stripes distinct, well contrasted, and regular. The selfs should be pure, and of a decided colour.

ASTER.

Two classes are generally provided for exhibiting the China (or German) Asters: (1.) Chrysanthemum-flowered, and (2.) Quilled. The former includes all those with flat florets, and the latter the varieties with quilled florets. Seed of the best strains should be sown in mild heat during April, in sandy soil and leaf-mould, and the seedlings pricked into boxes four inches apart, shaded for a time, and afterwards removed to a position near the glass in a cold frame. It is of importance that the plants receive no check from cold, crowding, or want of water, but that they should be strong, dwarf, and healthy for planting out in the end of May, or beginning of June. The ground on which they are to grow should be rich, and a little lime and soot forked into the surface previous to planting. Plant in rows two feet apart, and from one to two feet, according to the varieties, between the plants. They should be watered in dry weather and well fed, to produce large, fine flowers. The points of merit are: (1.) Size, form, and freshness of flowers; (2.) smoothness of outline and regularity of the petals; and (3.) colour, which should be pure or bright. The Quilled should be round and full, and the florets equal, regular, and free from coarseness.

CAMELLIA.

The Camellia as a cut bloom, although slightly lacking in gracefulness, has a chaste beauty and symmetry of form peculiarly its own. Some of the best varieties for exhibition are Adriana, Alba plena, Carlotta Papudoff, Comtesse, Lavinia Maggi, Cup of Beauty, Fimbriata, Lady Hume's Blush, Mathotiana, Princess Marie, Reine des Beautés, Targioni, and Thomas Moore. For the cultivation of the Camellia see "Camellia" in the First Division. The points of merit are: (1.) Size and form of blooms; (2.) arrangement and texture of the petals; and (3.) colour: if white, it should be pure; if coloured, dense and bright; if streaked, the colours should be distinct.

CARNATION.

The flowers of the Carnation are much admired for their beauty, perfume, and general usefulness in decorative work, and when well grown make a beautiful stand for exhibition. A few of the best for competition are Admiral Curzon, Florence Nightingale, Germania, Henry Cannell, James Douglas, Matador, Mr. Frederick William Bacon, Robert Halgrave, Rob Roy, Sarah Payne, Sybil, and William Skirving. For pot culture the plants should be potted in five-inch pots in October in sandy loam and leaf-mould, placed in a cold frame, and kept close and shaded till the new roots have started to grow. Air should then be gradually given, until the light can be withdrawn altogether, except during heavy rain and frost. During February they should be shifted into ten-inch pots for blooming, giving plenty of drainage, and using the same soil, with the addition of some old manure well rubbed down, replacing them in the cold frame, where they should remain until May. As the flower-stems shoot upwards they should be loosely tied to stakes, and as soon as the buds can be detected the best should be carefully preserved, and three only left on a plant. Weak liquid manure, made from soot and sheep-droppings, should now be given twice a week, which will cause the buds to swell and give a dark healthy hue to the foliage. As the flowers expand they should be slightly shaded, in order to preserve them; but this should not be overdone, else the colours will not be so bright. Carnations are also successfully cultivated in beds, but so much depends on external circumstances that this system cannot always be relied upon; and to ensure fine blooms, at least in northern parts, a number must be grown in pots. The points of merit are: (1.) Size and form; (2.) distinctness in the markings and colour; and (3.) substance of the petals and regularity in their disposition.

CHRYSANTHEMUM.

This flower stands higher in public estimation at the present time than probably any other in cultivation. The great improvement in form and colour, especially in the Japanese varieties, and the perfection to which cultivation has brought them, have no doubt been the means of raising them to their present popular eminence. Their culture engages the attention of a large number of horticulturists, but, even with all the knowledge at their command, comparatively few approach the ideal of perfection; and in treating of this subject we have endeavoured to point out the best known methods of obtaining good plants and perfect blooms. A selection of the best varieties for exhibition, as cut flowers, includes the foling :—

Incurved.—Alfred Salter, Barbara, Empress of India, Jeanne d'Arc, Golden Empress, J. Lambert, Lord Alcester, Miss M. H. Haggis, Mrs. Heale, Princess of Wales, Queen of England, and Violet Tomlin.

Japanese.—Avalanche, Boule d'Or, Carew Underwood, Edwin Molyneux, Etoile de Lyon, Jeanne Delaux, Madame C. Audiguier, Mdlle. Lacroix, Meg Merrilees, Ralph Brocklebank, Sunflower, Val d'Andorre, Viviand Morrel, and W. H. Lincoln.

Reflexed.—Annie Salter, Alice Bird, Chevalier Domage, Cullingfordii, Dr. Sharpe, Golden Christine, King of Crimsons, Mrs. Forsyth, Pink Christine, Peach Christine, and Purple King.

Anemones.—Acquisition, Fabian de Mediana, Fleur de Marie, Georges Sand, Gluck, Lady Margaret, Mdme. Theresa Clos, Mdlle. Cabrol, Miss Annie Lowe, Mrs. Pethers, Prince of Anemones, and Sœur Dorothée Souillé.

The culture of the Chrysanthemum is given in the First Division, and it is only necessary here to notice the special treatment required for the production of extra-large specimen blooms for exhibition as cut flowers. To attain this, strong, firm, healthy plants should be potted singly in eight-inch pots, giving the strong-growers one size larger, and using the same compost as has been recommended for bush plants. About the end of May the plants can be set in the open air, in rows at convenient distances apart, with strong posts fixed at the ends and wires stretched between them, to which the stakes supporting the plants should be firmly secured. Watering must be carefully attended to daily, giving it copiously where needed, and sprinkling the plants and ground between them late in the afternoon of dry warm days.

Taking the Buds.—Sometimes buds appear prematurely, when the top should be pinched off at a good leaf, from which fresh growth will be made, producing buds that will come in at the right time. Certain varieties require special treatment for the timing of the buds, such as the Rundle family, as representing the early varieties. These should be pinched from the middle to the end of May, which throws the first break into the end of June or beginning of July. The later varieties should be pinched between the last week in May and the middle of June, some allowance being made for the locality where the plants are grown. Cutting hard down is recommended at this time by some growers, but we see no advantage from this process further than reducing the height of the plants, which should not be considered by those growing for exhibition blooms, as the best have been and are obtained from plants which have been allowed to grow to their full height or to the crown buds. Regular attention should be given to tying the plants and rubbing out all side-shoots except the three leading ones produced at the first break, which takes place in June. The term "first break" is derived from a particular stage of the plant's growth. If the cutting has been rooted at the proper time and allowed to grow naturally with a single stem, a flower-bud will appear about the end of May or in June, the time varying with the variety and locality where it is grown. Immediately under this bud a number of small shoots will spring, three of which should be allowed to grow and the others rubbed off. During the month of August a bud will be formed on each of these shoots, which is termed the "crown bud." Shoots will also start below this bud, and they must be rubbed off as they appear. Should the crown bud, however, show too early, it should be promptly pinched out, and one of these shoots retained, and encouraged to grow as fast as possible. On it the second crown or "terminal bud" will soon appear, and must be "taken," or preserved, with every care, as it gives the bloom for exhibition. All other flower-buds must be ruthlessly rubbed off as soon as they appear, and so must all young shoots that may start from the stem. A little practice enables the operator to take the buds deftly, without injury to the chosen bud or to the stem on which it grows, and the best time to do it is in the early morning, or in the evening when the dew is on them. The crown bud is usually found the best for exhibition, and should be chosen when well timed. If the buds of late varieties are fit to be taken in the second week of August, it is the best time for them. The mid-season varieties come in best if taken from the middle to the end of August. The first fortnight in September is the most suitable time to take the early varieties. The flowers from the terminal bud are generally smaller and neater than from the crown bud, and should be chosen for certain varieties in both the Incurved and Japanese classes, which varieties should be observed and noted in regular practice. The largest flowers are obtained from plants with a single stem, and the first crown bud on the top of it. The plants should be gone over twice a week, to regulate the growth and carefully remove all useless side-shoots and suckers.

Mildew and Insects.—A constant watch requires to be kept for the first sign of mildew appearing on the plants, and at once dust the parts affected with flowers of sulphur. Aphides are also an insidious pest, and should be promptly kept in check by dusting the shoots on which they appear with tobacco powder. In some seasons earwigs are a troublesome nuisance, and they must be diligently watched and trapped whenever they appear; the hollow stalks of broad beans forming excellent traps when cut into lengths of 6 to 8 inches and placed among the plants and pots. They should be carefully examined daily, and the earwigs lodging in them are easily shaken or blown out into a

pail of hot water. Before housing the plants, it is a good plan
to give them a thorough syringing with a mixture of two
pounds of flowers of sulphur and one pound of soft soap to
twenty gallons of water; and after they are housed they ought
to receive a thorough fumigation, not too strong at one time,
but repeating it on several nights in succession, which will
clean them of aphides. When the leaf-mining grub is noticed
to be at work, by the pale grey or brown spots and streaks on
the leaves, these should be at once pressed between the finger
and thumb to kill the grubs, or if only a few leaves are
attacked, they should be picked off and burned.

Housing the Plants.—Some forethought is required here to
arrange the plants as far as possible to get them into flower at
the proper time. The end of September is the best time to
house the late varieties, leaving the early ones out for a week
or ten days longer, according to the weather. Those wanted
forward should be set in the warm end, and others too forward
in the cool end of the house. In order that the sudden change
from the open air to a close atmosphere may be felt as little as
possible, the house should be kept slightly moist at night, with
a free circulation of air about the plants until they get inured
to the changed conditions. When the buds begin to open, a
drier atmosphere should be maintained, and fire-heat may
be used if necessary for this purpose. Feeding with liquid
manure should still be continued, but only of moderate
strength, remembering that high feeding and a close moist
atmosphere are the agents which cause much of the damping
of the blooms. A slight shade from strong sunshine should
be given, so as to preserve the colour of the flowers. Artificial
heat must always be cautiously employed in pushing forward
the blooms; but Japanese varieties are improved by a little,
if it can be applied without risk of having them too forward.
Sudden changes of the atmosphere should be avoided, and no
liquid manure given after the blooms are about two-thirds
expanded. A bloom takes from three to four weeks to develop
itself fully after the opening of the bud, and the grower knows
whether to force or retard the plants to bring the flowers in at
the proper time. After the flowers are full blown, they keep

fresh for two weeks in a cool house if it is kept moderately dry and shaded.

Dressing the Blooms.—This is a necessary operation, and is practised more or less by all successful exhibitors. It consists in arranging, with suitable instruments, any florets that are out of place, extracting an eye, and pulling out any rough or damaged petals. This is best performed by the operator fastening the flower securely in the cup in which it is to be exhibited, and then deftly removing the eye and rough petals with a sharp-pointed knife and a suitable pair of tweezers. The florets should next be properly manipulated so as to look as even, neat, and perfect as possible. The Japanese do not require much dressing, but damaged florets and an eye when it appears should be removed, after which the bloom should be turned upside down and gently shaken, which is all it requires. Anemone-flowered should have badly formed disc florets taken out, and misplaced ray florets put into position. Reflexed flowers should also receive the needful dressing by removing imperfect and notched petals, and arranging the flower smoothly and shapely. All boards, cups, tubes, and boxes should be ready when wanted to save hurry at the last moment.

The chief points of merit are : (1.) Size ; (2.) symmetry ; and (3.) colour. Smoothness of outline and regularity in the disposition of the petals are imperative qualities in Incurved and Reflexed blooms. In the Japanese varieties the blooms should be large, full, and regular in outline, the length, breadth, and artistic curl in the petals being also important points in adjudicating on certain varieties.

DAHLIA.

This showy autumn-flowering plant has been long a special favourite with the florist, who has brought the flowers to a wonderful state of symmetry and perfection. The Show and Fancy Dahlias have been rather hard pressed for some years in public esteem by their beautiful rivals, the Single, Cactus, and Decorative varieties ; but popular taste is again leaning towards the old favourites ; and although they are not so useful for general purposes as the other sections, still they produce a splendid effect on an exhibition table, and command the notice and admiration of all true lovers of beauty and symmetry in flowers. The following selections of Fancy, Show, and Single varieties, including all colours, are among the best in their respective classes for the purpose of exhibition :—

> *Fancy Dahlias.*—Annie Richard, Flora Wyatt, George Barnes, Henry Glascock, James O'Brien, Lottie Eckford, Mrs. A. Holls, Mrs. Saunders, Polly Sandell, Prince Henry, Professor Fawcett, and Romeo.
>
> *Show Dahlias.*—Alexander Cramond, Clara, Emperor, Goldfinder, Henry Walton, Herbert Turner, James Huntley, J. N. Keynes, Julia Wyatt, Mrs. Gladstone, Mrs. Langtry, and Mrs. W. Slack.
>
> *Single Dahlias.*— Amos Perry, James Kelway, Miss Gladstone, Miss Maxton, Mrs. Bowman, Mrs. Fergusson, Mrs. H. M. Stewart, Terra-Cotta, T. S. Ware, Volunteer, White Paragon, and William Potter.

The proper cultivation of the Dahlia is of the greatest importance to the exhibitor, and requires a considerable amount of attention, perseverance, and skill to ensure success. To grow it for exhibition, a rather stiff loamy soil, richly manured, is the best. The ground must be well trenched the previous winter, at least two spits deep, at the same time giving it a liberal quantity of thoroughly made manure. About the end of May it should again be dug or forked over, breaking it well in the operation. Then lay it out in rows running north and south, with a distance of four to five feet between each, and not more than four feet between the plants. Stakes should be driven into the ground where the plants are to be put, to which they must be tied as they grow to prevent them being broken by the wind. In preparing the plants, the roots should be placed in heat to start about the end of April, giving them a temperature of 56°, and taking care not to allow them to get dry, which tends to throw them into premature bloom. This method of selecting fine plump roots and starting them with a single stem in preference to cuttings is adopted by competitors with the utmost success. When the shoots have grown about six inches, gradually harden them off, by admitting more air or by removing to a cooler place, guarding against any sudden check to the plants, and by the end of May they may be fully exposed to the open air. The first week in June is soon enough to plant them out, when all danger of frost is past. On no occasion should they suffer for want of water, while liquid manure twice a week will be beneficial. It should be used weak at first, but gradually stronger as the season advances. The result will be more satisfactory if the liquid manure is varied thus : farmyard manure at one watering, sheep-droppings at the next, guano or any other artificial manure capable of being dissolved at another, and so on. Three to five shoots is enough to leave on a plant to obtain the largest and finest blooms. A stake should be put to each shoot, and the laterals should be all pinched as they appear. At this stage a good mulching of decomposed manure will be of service to the health of the plants, especially if the weather is dry. About the beginning of August buds will appear in the points of the leading shoots, the best of which, or those of good form with full centres, should be selected and the others pinched off ; but if the first flowers are considered too early, they should be taken off, and lateral shoots trained up to take their place, which will produce flowers two weeks later. When the blooms begin to expand, they require protection, which can be given by fixing a piece of cardboard, to a stake, and placing it over the bloom. They should be cut when dry, and properly arranged before leaving for the show. The points of merit are : (1.) Size ; (2.) form ; (3.) colour. The flowers should be round, full in the centre, and not cupped. The petals of both double and single varieties should be regular and slightly reflexed, of good substance, and rich and clear in colour.

DELPHINIUM.

The bold handsome spikes of this hardy plant are held in high esteem by many cultivators of florist's flowers. There are both single and double varieties, with tints of colour from light blue to dark purple. Their tall spikes and attractive flowers are very much admired when well shown. The following are a few of the best varieties :—Bella Donna, Breckii, Desdemona, Mrs. Gamp, Prince Albert Victor, and Voltaire. The cultivation of the Delphinium is similar to that for the Antirrhinum, except that it is propagated by division, which is best performed in spring. Planted on a rich border, the plants will get thoroughly established the same season, and with a little attention during summer as to watering and mulching during dry weather, good spikes will be produced the following year. Plants may also be divided early in autumn, and kept in pots in cold frames through the winter; and if potted and planted out in March, spikes of fair quality can be obtained from them the same season. The points of merit are: (1.) Size of spike; (2.) size and number of expanded flowers; and (3.) quality of the flowers, which should be dense in colour, of good substance, and the eye distinct.

GERANIUM.

A very showy stand for exhibition is made from a good selection of Geranium flowers in trusses, and if properly set up and the colours well arranged, they produce a dazzling effect. The following are a dozen of the best, including varieties from each section :—Brilliant, Buffalo Bill, Caliban, Duchess of Portland, Freya, Gloire de France, Madame Thibaut, Opal, Rev. Dr. Morris, Silver Queen, Shirley Hibberd, and Charles Turner. The trusses should be exhibited in triplets, a thin wire being twisted down the stem, bending it slightly outwards to prevent the blooms from crushing. The points of merit are: (1.) Size and freshness of the trusses; (2.) size of pips and breadth of petals; and (3.) colour and texture of the flowers.

GLADIOLUS.

The Gladiolus never fails to be greatly admired at horticultural exhibitions, even by those who have no special knowledge of florist's flowers. Indeed it may be said that a collection of them, when well set up with most of the flowers expanded, has nothing to equal it for a gorgeous display in the cut-flower division. The following are some of the best exhibition varie-

ties :—Albatross, Albion, Bernard de Jussieu, Imperatrice
Eugenie, Leviathan, Lord Howard, Mont Blanc, Mrs. Derry,
Mrs. Laxton, Sir Percy Herbert, Reine Blanche, and Urania.
The Gladiolus is not so difficult to grow as many other florist's
flowers, but success will not be attained without considerable
skill and trouble. The ground ought to be trenched in the
autumn, giving at the same time a dressing of well-made
manure. The bulbs are best planted late in March or early
in April. The beds should be formed of four rows about
15 inches apart, and one foot between the bulbs, set three
inches deep. The plants should be carefully staked as they
grow, and a good mulching of manure given in dry weather.
Plenty of water should be supplied to the roots, and also liquid
manure as soon as the flower-buds begin to swell. As they
expand they must be shaded to procure first-rate spikes. The
most effective way to do this is to fix an oblong box with a
glass front to a stout stake, the box being of sufficient length
to contain the whole of the spike. As the lower blooms
expand the glass may be shaded with whitewash, always giving
the unexpanded blooms the full sun. In order to bring late
spikes forward for a show, they should be cut and placed in
bottles of warm water set in a hothouse. Those that are too
early may be cut, and kept in a cool shaded place to retard
them. When the flowers incline backwards, they should be
tied to the front until they are expanded, when they will remain
in that position. In arranging a stand, the spikes should be
set up with their tops slightly bent back and some of the
foliage put around them, which greatly improves their appearance. The points of merit are : (1.) Length of the spike ; (2.)
size and number of expanded flowers ; (3.) richness of the
colour and distinctness of the marking ; and (4.) texture of
the petals, the edges of which should be slightly reflexed to
show the interior of the flower.

GREENHOUSE FLOWERS.

A beautiful and interesting display is produced by a stand
of various kinds of greenhouse flowers, tastefully set up in
bunches, interspersed with a few sprays of graceful foliage.

The following are some of the best genera, excluding those provided with a class for themselves, such as Camellia and Rhododendron :—*Abutilon*, *Azalea*, *Bouvardia*, *Choisya*, *Clivia*, *Lapageria*, *Lisianthus*, *Lilium*, *Philesia*, *Plumbago*, *Statice*, and *Vallota*. The cultivation of these is given in the Plant Division. In fixing the number of bunches required on a stand in competition, *distinct genera* should be clearly specified in the schedule, else confusion often arises from the introduction of several species of the same genus, or, worse still, a number of varieties of the same species. Three or more, according to size, of the best-flowered shoots should be cut with long stalks, and wired, if necessary, to keep them in position. They should be lightly and tastefully arranged, using the foliage of each kind where possible with the flowers. The points of merit are: (1.) Arrangement; (2.) quality of the flowers; and (3.) freshness of the flowers and foliage.

HARDY HERBACEOUS PERENNIALS.

The popular taste has greatly revived in recent years for this useful class of hardy flowering plants, and their cultivation is rapidly extending as their excellent merits become better known. In the schedules of flower-shows they are generally stipulated to be exhibited as "spikes," practically excluding all other forms, and limiting the choice to flowers produced on spikes. This condition ought always to be modified so as to admit all forms of inflorescence—spike, raceme, umbel, corymb, fascicle, panicle, or cyme—by stipulating for so many "Single Stems in Flower of Hardy Herbaceous Perennial Plants." Many fine kinds, however, are more effective when set up in "bunches," and in such cases a certain number of stems in each bunch should not be exceeded, so that the competition may be on equally fair conditions.

Herbaceous plants are of easy cultivation, and will grow freely in any ordinary garden soil; but in the case of exhibition flowers, they well repay a little extra care and attention to their particular wants. As a rule, to get the finest exhibition specimens, the plants require to be divided every few years, planting the healthiest roots, or pieces, in fresh soil enriched with well-rotted manure. In dry weather attention must be given to watering and mulching them when necessary. The stems must be carefully staked, and the flowers protected from wind and rain as soon as they begin to open, so that they may be preserved in the best possible condition. In growing a collection for exhibition there is a wide choice of beautiful kinds, and the following eleven genera may be named as a good selection:—*Anemone japonica alba, Delphinium grandiflorum plenum, Dictamnus fraxinella, Doronicum columnæ, Gaillardia aristata, G. grandiflora, Inula glandulosa, Orchis foliosa, Phygelius capensis, Rudbeckia speciosa, Spiræa palmata,* and *Veratrum nigrum.*

The points of merit in herbaceous spikes are : (1.) The size, vigour, and freshness of the inflorescence ; (2.) size and quality of the flowers ; and (3.) the richness, contrast, and purity of the colours.

HOLLYHOCK.

This is of a stiffer and more robust character of spike than the Gladiolus; it also lacks its bright colours and graceful appearance. It has many votaries, however; and were it not so subject to disease, it would be better represented at exhibitions. The following are some of the finest varieties for competition :—Alba Superba, Charles Chater, Czar, F. G. Dougal, Her Majesty, Jane Wilson, Kerr's Scarlet, Mrs. Laing, Model, Pink Perfection, Queen of Buffs, and Queen of Yellows.

In the cultivation of the Hollyhock, the soil should be prepared as advised for Dahlias. A position sheltered from the wind should be chosen, but it must be fully exposed to the sun. As the Hollyhock is nearly hardy, it may be planted out any time in spring, while old plants may be left out all winter, with a slight protection in severe weather. Plants intended for competition should be struck from cuttings, or eyes, in the autumn and potted on as they require it. They should be grown in frames, protected only in severe frost, and planted out on the first favourable opportunity in spring. A liberal supply of liquid manure when they are growing is very beneficial. As the flowers on the lower part of the spike expand, they require shading, which is done by placing three stakes around the spike and enclosing it with thin canvas. This gives a longer spike of fully developed flowers. If wet weather sets in, the spikes should be so protected as to prevent injury to the flowers.

The points of merit of spikes are : (1.) Length of spike of expanded flowers; (2.) size and regularity of flowers, which should be close set but not overlap ; and (3.) the colours rich and clear. In single blooms the points are : (1.) They should be round, high in the centre, with the guard petals flat on the stand ; (2.) quality of petals ; and (3.) colour, which should be rich and clear.

ORCHID.

A stand of spikes or bunches of the finest Orchids, when properly set up, excels all others in elegance and richness of colour, and surpasses most kinds in striking appearance. There ought to be no limit to the exhibitor staging either a single spike or a bunch of spikes of the same variety, because the exhibit is meant to be attractive, as well as to show the results of cultural ability, and to satisfy the curiosity of the orchidophilist. A single spike of *Phalænopsis Schilleriana* is all that could be desired, while a single *Cypripedium* or *Lycaste* on the top of a stem has very little appearance ; but if these are wired and set up in a bunch, with a backing of green leaves, the effect is vastly more attractive, and therefore " bunches " should be the rule. The following are twelve of the best genera for this purpose :—*Aërides, Cattleya, Cælogyne, Cypripedium, Dendrobium, Lælia, Lycaste, Masdevallia, Odontoglossum, Oncidium, Phalænopsis,* and *Vanda.* The cultural directions are given in the Plants Division. The value of the exhibit depends much on the manner of arranging the bunches and setting up the spikes. The flowers should be carefully wired, so that the stalks may be bent into any position. They should be arranged in triangular form, with the flowers facing the sides as well as the front, and nothing is better than a few of their own leaves around them ; but if they are not to be had, a few fern fronds or asparagus sprays may be used instead. All leaves on the foot-stalks should be preserved, as they improve the effect, and although wires are used for setting up the flowers, the stems should be sufficiently long to reach the water, to keep the flowers fresh. The points of merit are : (1.) Arrangement and size of spikes and flowers; (2.) quality of the blooms; and (3.) colour, which should be rich, pure, and the marking well defined.

PANSY.

This is a special favourite with the amateur florist, around which his keenest interest centres, and its culture for exhibition is practically in his hands. Although lacking the showy appearance of some others, such as the Gladiolus and Dahlia, still it has many qualities peculiar to itself, which attract attention and enlist many horticulturists in its cultivation. The following are among the finest Fancy and Show Pansies for exhibition :—

Fancy Pansies.—A. M. Cocker, Bailie Donald, Beauty, David Henderson, John Pope, Lady Wolseley, Lord Rosebery, Maggie Bella, Miss Bliss, Mrs. John Downie, Neil M'Kay, and William Hayes.

Show Pansies.—Alexander Black, Countess of Rosebery, Devonia, John Elder, Lord Frederick Cavendish, Mauve Queen, Miss Philip, Miss Jessie Foot, Mrs. Gladstone, Mrs. Henderson, Sultan, and The Favourite.

The Pansy is propagated by cuttings in early autumn, selecting the shoots springing from the base of the plants, and inserting them in boxes containing loam, leaf-mould, and sand in equal proportions, placed in a cold frame and shaded till roots have been made. The young plants should remain in the frame till the following spring, when they may be planted out in March, if early flowers are wanted, otherwise in April, so that they may be established before the hot weather sets in. Pansies succeed best in new soil, which makes them less liable to die off during the hot months of June and July. Any good garden soil suits them, but a strong moist loam is the best. It should be well trenched and manured the year previous to planting, and cropped with potatoes, after which it will be in good condition for pansies. If exhibition blooms be required in early summer, a plantation must be made in autumn, and protected in winter from severe frosts and continued rains by placing a frame over it, the sashes of which can be moved at will. Three or four shoots will be sufficient for each plant, and they should be carefully staked to prevent them being broken. If flower-buds appear, they must be removed ; three weeks being the time necessary for the full expansion of the flower after the bud is seen. Slight shading is necessary for a few days before the flowers are gathered, shading the blooms only, which may be done by fastening a piece of cardboard to a short stake and placing the shade over the bloom. The plants should be kept well watered in dry weather, both at the roots and over the tops, while an occasional sprinkling of soot between the plants before watering will be beneficial. Weak liquid manure made from cowdung given once a week will greatly assist the perfect development of the blooms. An occasional syringing, when the plants are growing, with soap-suds will prevent greenfly attacking them ; but it must be discontinued when the blooms have been chosen. A light mulching with short grass prevents evaporation, and keeps rain from splashing the flowers. The flowers should be carefully handled when staging them ; the outer edges of the petals should be slightly reflexed, and touching the stand, which should be painted green, and no paper of any colour used beneath the flowers. The points of merit in a Show Pansy are : (1.) Size ; (2.) form ; and (3.) substance. According to the predominating colour they are divided into three classes, viz., selfs, white, and yellow grounds. The first comprises dark, light, and yellow selfs. These should be rich and pure in colour, with perfect centre. In the white and yellow classes the ground colour should be bright, clean, and distinct, the eye uniform and well defined. The Fancy Pansy may be said to comprise all others which have not the distinctive characteristics of the Show Pansy. They should be large, circular, petals full, and overlapping.

PENTSTEMON.

The Pentstemon is a general favourite with exhibitors, from the fact that it is very accommodating, easily grown, and can be exhibited in good form with very little trouble, provided that the proper varieties are cultivated. Among the best for exhibition are Eclipse, F. W. Moor, Mont Blanc, Renown, T. S. Ware, and W. E. Gladstone. Their cultivation is exactly similar to that recommended for Antirrhinum. The points of merit are : (1.) Size of spike and number of flowers ; (2.) size and texture of the tube, the lip of which should be slightly extended, and turned back to show the throat ; and (3.) colour, which should be dense and clear, and the markings well defined.

PHLOX.

This is a very showy flower, and when represented by good varieties, well cultivated, they make a fine display. Some of the best are Burns, Captain R. Jackson, Earl of Mar, John Stewart, Miss Lingard, and Panama. The cultivation is similar to the Antirrhinum. The points of merit are : (1.) Size of spike and flowers ; (2.) form and colour. Selfs should be pure, of decided colour, the markings uniform, and clearly defined.

PICOTEE.

A very useful and pretty hardy flower, which is deservedly popular at exhibitions. Twelve of the best Picotees for exhibition are Ann Lord, Brunette, Evelyn, Favourite, Hero, Lady Holmesdale, Mrs. Gordon, Mrs. Sharp, Paulina, Thomas Williams, William Summers, and Violet Douglas. Their cultivation is similar in every respect to that of the Carnation in the open air. The points of merit are : (1.) Size and form of the bloom ; (2.) substance and regularity in the disposition of the petals ; (3.) colour, which should be pure and bright. The lacing may be light or heavy, but distinct.

PINK.

This is another of the very useful hardy plants, which should receive more attention on account of its fragrance and free flowering nature. Among the best for exhibition are John Love, J. Melville, Mrs. Sinkins, Mrs. T. M'Crorie, Mrs. Welsh, and Robert Holmes. The cultivation is similar to the Carnation. The points of merit are: (1.) Size and form of the flower; (2.) colour and texture of the petals; (3.) markings, which should be distinct and well defined. If of one colour, it should be pure and decided.

PYRETHRUM.

A class of useful hardy herbaceous perennials, which has been greatly improved in recent years, and become a general favourite. A stand of well-grown blooms, set up with their own pretty foliage, almost rivals the best Chrysanthemums. A few of the best for exhibition of the double and single varieties are A. M. Kelway, Aphrodite, Cleopatra, Comet, Duke of Edinburgh, Chamois, John Holborn, L. Kelway, Magician, Melton, Model, and Sylphide.

The Pyrethrum is very easily cultivated, growing freely in any good well-manured garden soil. The ground should be trenched two feet deep, and the plants put thirty inches apart, and well watered and mulched in dry weather. Very little is required until the buds appear, when they should be thinned out, and the best encouraged to grow strong by applications of liquid manure once a week. The flowers should be protected from the sun and rain after they are expanded. The points of merit are: (1.) Size; (2.) form; (3.) colour. The flowers should be round and full in the centre, and the petals free from all blemishes. If the colour is white, it should be pure, and bright if coloured.

RHODODENDRON.

The flowers of the Rhododendron make a fine exhibit when they are done up with leaves and neatly arranged in a stand. Fine large trusses set up singly have a splendid appearance, and should always be exhibited in that manner. The following are some of the best, including both greenhouse and hardy varieties:—Album grandiflorum, Charles Bagley, Countess of Haddington, Dalhousianum, Edgworthii, Lobbii, M'Nabii,

K

Melton, Mrs. John Clutton, Sir Robert Peel, Sir Wm. Armstrong, and Vestal. The points of merit are: (1.) Size of truss; (2.) size and texture of the flowers; and (3.) brightness of the colours and distinctness of the markings.

ROSE.

Although the advance in quality has been less conspicuous of late in the Rose than in almost any other class of florist's flowers, still it nobly holds its own, and maintains as worthily as ever its proud title of " The Queen of Flowers "—combining beauty of form, richness and diversity of colour, and exquisite fragrance with great adaptability for decorative purposes, and especially for making the most lovely of all displays at an exhibition. The principal classes cultivated for exhibition are the Hybrid Perpetual and the Tea Roses. The following selections of a dozen of each includes the best of the different shades of colour, so that the stand may be as effective as possible :—

> *Hybrid Perpetual Roses.*—Baroness Rothschild, Comtesse de Chabrillant, Duchess of Bedford, Duke of Edinburgh, Etienne Levet, Horace Vernet, La France, Lord Beaconsfield, Mabel Morrison, Madame Victor Verdier, Marie Baumann, and Xavier Olibo.
>
> *Tea-Roses.*—Anna Olivier, Catherine Mermet, Cleopatra, Madame Bravy, Madame de Sirtot, Madame de Watteville, Madame Pierre Guillot, Mrs. James Wilson, Niphetos, Princess Beatrice, Souvenir d'Elise, Vardon, and The Bride.

These are mostly old and well-known varieties, but their excellent qualities are not easily rivalled by any others of their class and colour.

To grow the rose to perfection, a sheltered situation, fully exposed to the sun, is best, and a rich, strong, loamy soil is required. If the soil is naturally light, it should be made more suitable by adding strong loam to it, the top spit of an old pasture being the best when it can be procured. The ground should be trenched and made up of good sound materials to a depth of about two feet, incorporating plenty of well-rotted manure, wood-ashes, or other enriching material as the work goes on. To obtain the largest and finest blooms, dwarf low-budded plants of Hybrid Perpetuals on the Manetti stock should be planted in the month of November. The planting should be carefully done, the long bare roots neatly cut in, and the fibres evenly spread out as near the surface as possible, so that the budded part is fairly covered with the soil. The plants should be set about three feet apart, secured to neat stakes, and the surface of the soil kept well mulched with stable litter. The unripe tops of the shoots may be shortened when planted, but pruning should be delayed till the middle of March, and then the shoots reduced to two, with three or four eyes in each. No exhibition blooms may be expected the first year, but everything should be done to promote vigour and maturity in the plants for the production of first-class blooms in the second year, when roses are generally at their best. Only two shoots from a root should be allowed to grow. They should be carefully attended to, kept clean, and all the flower-buds picked off as they appear, so as to get the wood as strong and well ripened as possible. In March of the second year the shoots should be pruned again to about six inches from their base, and two young shoots allowed to grow from each, and upon these the roses for exhibition are grown. As soon as the best buds can be clearly seen, disbudding should commence, carefully reserving the best formed and plumpest buds that promise to open their flowers at the date of the exhibition. Three buds on each shoot are enough at a time, and all others must be remorselessly picked off, removing also all superfluous growth, so that the whole resources of the plant be thrown into the development of the exhibition roses. Water must be given in dry weather, and liquid manure liberally applied when the buds are set and swelling off, keeping all clean from insects and mildew.

The Tea-Roses thrive well under similar conditions as to soil and treatment; they are, however, not so hardy as the Hybrid Perpetuals, and require to be well protected with a deep covering of litter during the winter. Even with protection the wood is frequently killed to the ground, but the litter saves the frost getting at the roots, and most varieties will start again freely from the collar. The Teas do best when budded on the briar stock; they should not be so close pruned as the Hybrid Perpetuals, as they produce the finest blooms

from about the middle of the previous year's wood, when it is
well matured.   The points of merit in Roses are : (1.) Size ;
(2.) form ; and (3.) colour of the flowers.   The larger the bloom
the better, provided it is full in the centre and of a good shape.
The petals should be thick, glossy, and regularly disposed, free
from all blemishes and notched edges, and the colour should
be clear and fresh.   The blooms should always be exhibited
with foliage, which gives them a fresh and finished appearance,
and where it is necessary to add foliage it should not be overdone, good taste easily determining the right proportion.

STOCK.

There are several kinds of Stocks (*Matthiola*), which among
them furnish a succession of flowers from early in the season
till late in autumn.   Sown the previous year and treated as a
biennial, the Brompton Stock begins the season, followed by
the Ten Weeks' Stock, an annual the name of which indicates
its precocity ; after which the Intermediate or East Lothian
Stock carries on the supply as long as mild weather lasts.
Seed should be saved from only the very best strains ; and for
producing double flowers, the plants from which the seed is
saved should be as old as it is possible to keep them before
allowing them to bear seed.   The seed of Intermediate and
Ten Weeks' Stocks should be sown at the same time, and
the plants treated in the same manner as Asters.   Stocks
should be exhibited either in a " single spike "—the proper
way as *cut flowers*—or the whole plant should be set up, with
or without the root ; but whatever form is adopted, it should
be clearly stipulated in the prize schedule to prevent confusion.
The points of merit are : (1.) Length and diameter of the
spike ; (2.) number and quality of the flowers, which should be
perfectly double ; and (3.) substance of the petals, of which the
colour should be clear and bright.

STOVE FLOWERS.

This class is so profuse in excellent kinds of cut flowers for
exhibiting in bunches, that it is rather difficult, in making a
selection, to know exactly which to leave out.   However, the

line must be drawn somewhere, and the following genera will be found among the best and most suitable for the purpose :— *Allamanda, Amaryllis, Anthurium, Dipladenia, Eucharis, Franciscea, Gardenia, Ixora, Pancratium, Poinsettia, Rondeletia,* and *Stephanotis.* Their cultivation is given in the First Division. The manner of arranging the bunches and staging them is the same as is recommended for Greenhouse Flowers, and the points of merit are also similar.

VIOLA.

The Viola has deservedly come into popular favour very much of late years, owing to the great improvement in its varieties. It is one of the hardiest and most useful of plants, and when the flowers are set up on wires in bunches, they make a pretty display on an exhibition table. The following varieties are among the finest for competition :—Archibald Grant, Buccleuch Gem, Countess of Hopetoun, Countess of Kintore, Duchess of Albany, Evening Star, Gipsy Bride, Holyrood, Lord Elcho, Mrs. John Clark, Skylark, and York and Lancaster. The Viola is cultivated in a similar manner to the Pansy. If planted in September, they have sufficient time to establish themselves before winter, and are less liable to be thrown out of the ground by frost. When planted in autumn they commence blooming in early spring, and continue throughout the season, but for very late flowers they are best planted in the spring. The points of merit are : (1.) Size and form of bloom ; (2.) texture of petals ; and (3.) distinctness and brightness of colouring.

BOUQUETS AND FLORAL ARRANGEMENTS.

When this section of an exhibition is well filled, it never fails to attract the attention and admiration of all classes of visitors. They naturally look to this department for a display of good taste and artistic arrangement, and for enlightenment on the best methods of using cut flowers with the most perfect effect in a style of decoration which has in recent times become very fashionable. Therefore all illustrations and information tending to efficiency in the art of floral decoration are welcomed by the public.

A HAND-BOUQUET should be built as light and airy as possible, in shape like two-thirds of a globe, and about nine inches in diameter of flowers, exclusive of the fringe of fern fronds or other greenery. In competition bouquets the dominant colour of the flowers should be white, with a due blending of delicate shades of pink, orange, and red, all of which should harmonise with each other, and be tastefully set off with light graceful sprays of maidenhair fern. All violent contrasts are objectionable, and gaudy colours should be carefully avoided. The choicest and finest flowers at command should be employed, and every bloom selected and examined with care to see that it is perfect. Some of the best flowers for bouquets are those of the Bouvardia, Carnation, Chrysanthemum, Erica, Eucharis, Gardenia, Lapageria, Lily, Lily of the Valley, Narcissus, Orange, Orchids, Pancratium, Pink, Rose, Stephanotis, Tabernæmontana, and Tuberose.

No frames should be used or allowed in the building of a bouquet, but all the flowers should be carefully and neatly wired. A piece moist wadding should be placed on the cut end of the stem with the wire round it, thus affording a little moisture, to retain the freshness of the flower as long as possible. In commencing to make a bouquet, a groundwork of flattish flowers should be formed about six inches wide, into which the others should be regularly and tastefully inserted, keeping the upper edge of the petals about two inches above it without showing the wires. Each truss or bloom should stand clear by itself, the various colours and forms being distributed over the bouquet in harmonious order, and far enough apart to allow the groundwork to be seen between them. A few sprays of *Adiantum gracillimum* should next be inserted, spreading them gracefully over the top, and neatly finishing off with a fringe of *Adiantum cuneatum* fronds carefully placed round the edge. Bouquets built in this style being free and open, look more natural and graceful than with a close plain surface. Tastefully designed bouquet-papers should be used for all hand-bouquets, and should be put on when staging them.

The BRIDE'S BOUQUET should be of the same size and build

as the ordinary hand-bouquet, but composed entirely of white flowers; the sweet-scented orange-blossom being indispensable, with the usual setting of graceful sprays of maidenhair and sprigs of myrtle.

A TABLE BOUQUET should be about 12 inches in diameter, and built on the same principle as the hand-bouquet. In making a table-bouquet for exhibition, a choice selection of free sprays of large brilliant-coloured flowers may be used along with those already mentioned, such as Allamanda, Amaryllis, Azalea, Camellia, Dipladenia, Hibiscus, Lilium, and Rhododendron. Graceful sprays of fine foliage plants can be employed with good effect when placed tastefully among the flowers and around the edge. Long fronds of choice maidenhair fern should be gracefully disposed all over, and a tasteful fringe of greenery round the edge to finish it off. The flowers may be wired to keep them in position, but a bouquet-paper should not be used.

BUTTONHOLE BOUQUETS AND LADIES' SPRAYS.—The former should be about three inches high and two inches wide at the broadest part. The latter about nine inches in length and three inches wide. They require to be made with great taste and neatness, and of the choicest flowers. They should be arranged on wires, beginning with a small graceful point, gradually widening, and quickly rounding off to the stem. They are generally exhibited in sets of six, and may be all made of different flowers or in pairs. The smaller kinds of flowers should be used, such as Orchids, Carnation, Bouvardia, Jasmine, Lily of the Valley, Forget-me-not, and Rosebuds. The natural leaves of the flowers should be used along with neat fern fronds as a background.

A FLORAL WREATH should be about 18 inches across, and composed of choice flowers, which may be pure white or coloured, according to taste. It should be built on a galvanised frame, or ring of strong wire, covered with moss, over which should be placed sprays of Cypress, Ferns, Mahonia, or such like, to form a background. The flowers should all be wired (using No. 20) to go through the moss and keep them in position. A groundwork of flowers should first be made over the sprays, and then the largest flowers placed in the centre, with the smaller ones disposed along the sides, and ferns tastefully spread over all, giving an airy appearance to the whole.

EPERGNES.—These are of various shapes and sizes, but the most graceful form should be employed, because it is impossible, even with the best of skill and material, to produce the same effective arrangement on one of a clumsy make as could be done with one of a graceful design. An excellent design is one with a shallow cupped base and a trumpet vase $2\frac{1}{2}$ feet high, with three arms bending gracefully out from the stem, and to the ends of which are attached small hanging baskets. In filling the cupped base, a little clean green moss should be placed in the water, to rest the stems of the flowers upon to keep them in position. The top or trumpet should be filled with slender graceful flowers, the baskets with the finest specimens, and the base with a mixture of large flowers and foliage, with fern fronds spread over, covering the edge of the dish, the points resting on the table. Trailing pieces of Selaginellas or *Lygodium scandens* should be neatly suspended from the trumpet, and buds of Fuchsia or Begonia drooping over the edge of the baskets prove very effective.

In judging this class, the peculiarities of taste will always play a part, whether right or wrong, in arriving at a decision, but with the size clearly defined and the style of arrangement indicated, exhibitors will be enabled to enter into the contest on even terms, and the chances of an unsatisfactory verdict be reduced to a minimum. The points of merit are : (1.) Taste displayed in the arrangement; (2.) quality; and (3.) variety in the flowers,

DIVISION III.

FRUIT.

THIS is always an important department in horticultural exhibitions, and more particularly in the autumn, when the tables are laden with the richest treasures of Pomona, culled from the hothouse, garden, and orchard. In no other section is there such keen and persistent competition, and the laudable ambition of every good grower of fruit to win high honours at first-class shows is a marked feature of the times. The great enthusiasm displayed, and the long anxiety and hard toil cheerfully undergone by most cultivators who reap the laurels in fruit competitions, are not excelled, if equalled, in all other branches of gardening. The ever-increasing demand for the very best fruit of every kind has led to a considerable revival of the public interest in home-grown fruits, and the culture of hardy fruits has received a great impetus during the past decade, owing partly to the unremunerative nature of agricultural crops and the desire to find something more profitable than the growing of grain, and partly to the interest which has been stirred up in the question by a series of Conferences on Fruits, held at Chiswick since 1883 by the Royal Horticultural Society of London, and at Edinburgh since 1885 by the Royal Caledonian Horticultural Society. The former Society has held Conferences and meetings on Apples, Pears, Grapes, and other Fruits, at which their nomenclature was rectified, the best varieties named, and the methods of culture discussed, all of which has been embodied in the Reports of the meetings issued by the Society, and form one of the most valuable of pomological records in existence. The Scottish Society held a Congress on Apples and Pears in 1885, and on Plums in 1889, and the mass of practical information of the

most useful nature to the cultivator of those important Hardy
Fruits, which was gathered in connection with them, was published in the Reports issued by the Society.   The proceed-
ings at those Conferences and the Reports, and the comments
thereon in the daily papers and the horticultural press, have
awakened a deep interest in the public mind as to the great
importance of the adoption of improved methods of cultivation
and the extension of fruit-growing for market, as a profitable
branch of rural industry and a means of increasing the home
supply of wholesome food for the people.   It is therefore in
accordance with the nature of things that the fruit section in
horticultural exhibitions should be a centre of great attraction
to the public as well as of the keenest competition amongst
growers of fruits.

THE APPLE.

The Apple (*Pyrus Malus*) is a native of Britain and most
European countries, and its cultivated forms have spread to all
the civilised parts of the temperate zones, thriving as well in
Southern Tasmania as it does in Northern Canada, and in
both producing as fine fruit as the best that can be raised
in its native habitats in Europe.   It is supposed to have
been cultivated in Britain by the Romans, and since the
Middle Ages at least it has been the chief fruit grown in this
country.   Many fine showy kinds are now imported in large
quantities from abroad, and the dessert varieties are, as a
rule, sweeter and more melting than the same varieties of
home growth, but they lack the fine aroma and piquant
flavour of the latter, when in well-ripened condition.   The
culinary varieties from abroad are also large and very handsome in appearance, and find a ready market among our
teeming population, but they are much inferior to the best
home-grown culinary varieties in brisk, sprightly flavour, and
good cooking qualities.

Apples at exhibitions are usually divided into two classes,
Dessert and Culinary varieties.   Of the former the following are a few which ripen early, and are fit for exhibition at autumn shows as "ripe" dessert fruit :—Devonshire

Quarrenden, Irish Peach, Kerry Pippin, Oslin, Thorle, and *Worcester Pearmain. These all succeed and bear freely on any form of a tree, but for growing them to supply superior fruit for exhibition, the " bush " form of tree on the Paradise stock is the most convenient and generally the best. When properly attended to, by careful summer pinching, replanting and root-pruning when needful, top-dressing, and watering with liquid manure, such trees produce the very finest samples of fruit in abundance ; the crop requiring to be well thinned to allow the best fruits to attain perfection.

For a general collection of dessert apples to be exhibited " ripe or unripe," the following are among the best and most useful varieties in addition to the above six :—Adam's Pearmain, *Beauty of Bath, *Blenheim Pippin, Braddick's Nonpareil, Claygate Pearmain, Cockle's Pippin, Cornish Aromatic, Cornish Gilliflower, Court of Wick, Court Pendu Plat, *Cox's Orange Pippin, *Duchess of Oldenburg, Dutch Mignonne, Golden Pippin, *Gravenstein, *King Harry, *King of the Pippins, *Lady Sudeley, Mannington's Pearmain, Margil, *Red Astrachan, Reinette du Canada, *Ribston Pippin, and *Scarlet Nonpareil.

Among the finest varieties of culinary apples for exhibition in a collection of " ripe or unripe " fruits the following thirty may be named :—Alfriston, Annie Elizabeth, *Beauty of Kent, Bedfordshire Foundling, *Blenheim Pippin, Bramley's Seedling, Cox's Pomona, *Ecklinville Pippen, *Emperor Alexander, Frogmore Prolific, *Gloria Mundi, *Golden Noble, Grenadier, Kentish Fillbasket, Loddington, Lord Derby, *Lord Suffield, *Mère de Menage, *Mrs. Barron, *Peasgood's Nonsuch, *Pott's Seedling, Prince Albert, Red Winter Reinette, Stirling Castle, Striped Beaufin, Tower of Glammis, *Warner's King, Wellington, Wilts Defiance, and Winter Hawthornden.

The finest twelve varieties of dessert and culinary apples for exhibition are marked with an *asterisk* in the above lists. Their large and handsome appearance, and the fine rich colour of most of them, make them very attractive dishes on an exhibition table. Many other handsome and useful apples are extensively grown in various parts of the country, and some of them may take precedence, when well grown, over those in the lists, which are only intended as a guide to the uninitiated in selecting the varieties for successful competition. All exhibitors of fruit select the best varieties at their command, with little regard as to whether they have only a local or a widely known reputation.

Well-formed, healthy, dwarf, pyramid trees, grafted on the proper stock, should be obtained early in Autumn, or as soon as the leaves have fallen, and be planted on a well-prepared loam, spreading out the roots, and keeping them near the surface. Provision should be made for the production of early fruit, as well as for the growth of the more tender varieties, by planting them against a wall. After planting the trees, and staking those that require it, a mulching of stable manure should be spread over the roots. Defer pruning till spring, when the leader should be shortened to about one foot in length, and the side-shoots to about nine inches. As the young shoots grow they should be pinched when about six inches long, to form fruit-bearing spurs, continuing this operation as long as growth is being made, and then very little winter pruning will be necessary.

The apple is subject to the attack of many insects, but the caterpillar of the winter moth is the most destructive. Pinch the infected leaves and growths to clean the trees of insects, and remove all small and deformed fruit after they are set and swelling. Not more than one fruit on each spur should be left at the final thinning, as overcropping invariably ends in failure. Liquid manure should be freely applied when the fruit is swelling, and the ground mulched to produce a healthy root action and large fruit. It will not do for the exhibitor to neglect the trees all summer, and go round them the day before the show looking for the apples that will bring him success. He must diligently look to their wants from the time they start into growth until the fruit is ready to gather. The trees should be root-pruned if necessary as soon as the leaves begin to fall in October, by making an excavation about three feet from the stem, to admit of all downward running roots being cut; which operation has also the tendency to prevent canker in the tree, and to promote the formation of fruit in place of wood buds. To grow first-rate fruit of Cox's Orange and Ribston Pippin, as well as many other fine varieties, a wall is necessary outside of the most favourable localities. The trees should be trained on the cordon system, by leading two or more shoots up the wall. By pinching in the side-shoots to form spurs, keeping the trees clean, and feeding the roots while the crop is swelling, excellent apples should be obtained. The points of merit are: (1.) Large size; (2.) handsome form; and (3.) skin, smooth and clear in colour. Dessert apples should be sweet, juicy, and melting, with a rich pleasing aroma. Culinary apples should be acid, crisp, and juicy, and, when cooked, the flesh should be clear in colour and melting, with a brisk pleasing acidity.

THE APRICOT.

The Apricot is considered to be indigenous to Armenia, but has been found growing wild in many parts of Asia and Africa with a semi-tropical climate, where in most instances, however, it is supposed to be an escape from cultivation. It is stated to have been introduced from Italy to Britain in 1524 by Woolf, gardener to Henry VIII.; and in this country it requires a wall and a well-sheltered situation to bring the fruit to perfection. A well-built brick wall is the most suitable for growing the Apricot; and in favourable situations it may face any point of south between south-east and south-west, but in cold districts it must face due south. Some of the best exhibition varieties are Hemskirke, Large Early, Moorpark, and Shipley's. The border on which they are grown should be well drained, as the Apricot is impatient of stagnant moisture. Like most stone fruits, it delights in a calcareous soil, and if lime is absent from the fibry loam with which the border is formed, some lime rubbish should be added, and also a little sharp sand and wood-ashes if the soil is of an adhesive nature. Young trained trees, healthy and well ripened, should be planted as soon as the leaves have fallen, watering them if the soil is dry, and mulching to prevent evaporation and to protect the roots in winter. By careful pinch-

ing, disbudding, and regulating the shoots during summer, the
trees should, in two or three years, be in a fit condition for
the production of first-rate fruit, suitable for exhibition. In
the summer treatment, the shoots should be kept equal in
strength, because strong gross wood is very liable to disease,
and to cause disfigurement of the tree. To prevent this,
pinch and depress the strong shoots ; and if gross growth
still continues, or if canker appears, lift the trees, prune
the roots, and replant in fresh soil. The principal points
in the successful cultivation of the Apricot are : the protection of the blossom ; thinning of the fruit, and giving
it clear room to swell to its greatest size ; thinning and
training of the young shoots to facilitate their ripening,
as they produce the finest fruit ; careful watering in dry
weather ; giving liquid manure when the crop is swelling ; and
mulching heavily to keep the border cool and moist in hot
weather. Although the Apricot is comparatively free from
the attack of insects, red-spider and greenfly sometimes make
their appearance, but they are easily got rid of by the usual
means ; and if woodlice attack the fruit when it is ripening,
they should be at once trapped. The fruit should be gathered
before it is over-ripe, and placed on soft paper in the fruit-
room till it is wanted, when it should be wrapped in tissue-
paper and cotton-wool, and carefully packed for transit to the
exhibition. Nine fruits make a nice dish, and in setting them
up, a piece of cotton-wool should be placed in the centre of
the dish and covered with leaves, on which the fruits should
be equally disposed without actually touching each other. The
points of merit are : (1.) Size ; (2.) flavour ; and (3.) colour.

THE BANANA.

This is one of those rare tropical fruits which in cultivation
combine the useful with the beautiful in a marked degree.
Its culture as a handsome foliage plant has been already
described in the First Division. *Musa Cavendishii* also furnishes the exhibitor with a first-class dish of fruit for competition, to secure which it is necessary that it be planted in a

rich and well-prepared border, made of the same materials and receiving the treatment as described for foliage plants, till the fruiting stage is reached, or when the plant has attained full size. Water should then be withheld for about a month, and with a thorough soaking of tepid water applied to the roots at the end of that period, the plant will immediately throw up a bunch of fruit. A high and moderately moist temperature must be maintained to set and develop as many "hands" of fruit as possible; and when no more will set, the point of the bunch should be cut away. Abundance of heat, moisture, light, and air, with copious drenchings of liquid manure at the roots, will swell the fruit to its finest size. As it approaches maturity, or about ten weeks after flowering, water must be gradually reduced, and entirely withheld as soon as the first fruits show colour, with the atmosphere kept as dry as possible to prevent the fruit becoming spotted. As soon as they are all about an equal size and the bunch about half-coloured, it should be cut and hung up in a dry warm place and shaded from bright sunshine, when it will assume a rich yellow colour throughout, in a few days, and is then in a proper state for exhibition. A well-grown and properly timed plant will supply a bunch of fine ripe fruit within a week of a given date under careful treatment. The points of merit are: (1.) Size of bunch; (2.) size, equality, and ripeness of the fruit; and (3.) colour, which should be a clear yellow.

THE CHERRY.

The Cherry has been cultivated in Britain for its fruit for many ages, and in some parts, particularly in Kent, cherry orchards form an important and lucrative branch of rural industry. As an exhibition fruit it is held in considerable esteem, especially early in the season, when other fruits of high quality and attractive appearance are not plentiful. It ripens first of all our hardy tree fruits, and when grown under glass with very little artificial heat, it attains great perfection in May and June, and is then a highly appreciated and much valued dish of dessert. Among the finest varieties for exhibition are the following :—(1.) Dark-coloured : Black Tartarian, May Duke, and Morello ; (2.) Light-coloured : Bigarreau Napoleon, Elton, and Frogmore Early. The Cherry prefers a deep, well-drained, friable loam, and thrives well under the same treatment as other hardy fruit trees. It bears freely on bushes and standards in favourable situations, but in most parts of the country it requires to be trained against a good wall to obtain exhibition fruit of the highest excellence. The finest fruit is obtained by paying due attention to keeping the trees healthy and vigorous, free from all insects, pinching and regulating the shoots, thinning the fruit to a moderate crop, and securely protecting it from the ravages of birds, mulching the surface to protect and encourage the roots, and a judicious use of liquid manure to swell the fruit to its largest size. When nearing maturity the fruit should be fully exposed to the light, to colour and flavour it to perfection. The Cherry makes a beautiful dish when built in the form of a cone with all the stalks hidden from view, and in a collection of hardy fruit it is almost indispensable. The points of merit are : (1.) Size ; (2.) flavour ; and (3.) colour and equality.

THE CURRANT.

The Currant is one of our most useful and prolific bush fruits. Both the Black and Red species are indigenous to Britain, where they are extensively grown and used in a ripe state for a variety of purposes, beside being among the best fruits for the manufacture of preserves. The fruit keeps fresh and good for weeks of ordinary weather on the bushes after ripening when protected from the ravages of birds, and forms an attractive dish for exhibition. Being cultivated by every one who has a garden, a keen contest for supremacy on the exhibition table is engaged in, especially at country shows, and although few reach the standard of perfection, the majority of dishes are creditably shown. When Currants are well treated they can be grown to a large size, and the cultivation of large berries should be aimed at, whether they are intended for exhibition, market, or private use. Among the best Black varieties are Black Champion and Black Naples; Red varieties, Defiance and Warner's Grape; and White varieties, Cut-leaved White Dutch and White Versailles. Young plants are easily raised from cuttings, and in two years are fit for planting. Good fruit can be grown on bushes in the open, but walls are best for growing competition fruit—a south wall for the earliest, and a north aspect for later use. By planting on different aspects a supply can be had from midsummer till late in autumn. The shoots of the Black Currant should be thinly laid in and trained over the wall. In training the Red and White Currants, four shoots should be led up the wall at 15 inches apart, and the side-shoots kept pinched in summer and pruned in winter to about two eyes from their base. Plenty of light and air should be admitted to the fruit by pinching and cutting out superfluous wood, and the clusters thinned if too numerous. In June a top-dressing of well-rotted poultry-yard manure mixed with fresh loam and soot is beneficial to all Currants, and liquid manure, applied as the fruit is swelling, improves the size of bunch and berry. Black Currants should be gathered and exhibited singly, with part of the footstalk attached to the berry, while the Red and White should be exhibited in the bunch. They make a nice dish when arranged with the stalks in the centre out of sight, and neatly finished off in a conical form. The points of merit in Black Currants are: (1.) Size; (2.) flavour; and (3.) colour. In the Red and White: (1.) Size and flavour of berries; (2.) number and equality of the berries in the bunch; and (3.) colour.

THE FIG.

The Fig is one of the most fruitful of trees, some varieties producing, under favourable circumstances, three crops in the year. It is a native of Europe and certain parts of Africa, and it lives to a great age. Good crops of fine fruit have been grown on trees against south walls, in favourable districts, as far north as Ross-shire, with some protection in winter and spring, but they cannot be depended upon except in the best localities in the South of England. Among the best exhibition

L

varieties are Brown Turkey, Castle Kennedy, and Negro Largo. Where a house is devoted to the Fig, its culture is a very simple matter; but as a fig-house is only found, as a rule, in extensive gardens, a few hints on its cultivation among other plants may be of service. It thrives and fruits well on the back-wall of a vinery when not too much shaded, if the roots are confined to a limited space among rough loam, lime rubbish, brickbats, and some half-inch bones. For the first crop of figs ripe in May and June, a vinery started in January is necessary, the fruit being borne on the previous year's growth. If the crop is too thick, it should be thinned, leaving the best fruit at six inches apart on the shoots. As soon as they begin to swell, tepid liquid manure should be freely given, and the syringe applied morning and evening until the fruit begins to ripen, when the moisture should be reduced, and only sufficient water given as will keep the plants healthy. Figs in pots, after being well rested, should be introduced to the second vinery early in March. They should be potted in the same material, and, if well supplied with liquid manure when the crop is swelling, good exhibition fruit should be ripe from June till August. A supply from then till October will be furnished by trees in the cool peach-house, grown under the same conditions as those on the back wall of the vinery. The second crop from trees in the vinery should also be available to meet the demand for autumn shows. The Fig is a very gross feeder, and it is by the judicious application of liquid and solid manures that the finest fruit is grown. The fruit when ripe should be cut close to the stem with a sharp knife, to prevent injury in gathering, then placed in tissue-paper and packed carefully for transit. Six large fruits are sufficient to make a dish. The points of merit are : (1.) Size ; (2.) colour ; (3.) flavour ; the skin should be dry, and not too much burst open.

THE GOOSEBERRY.

The Gooseberry belongs to the genus *Ribes*, and is a native of Britain and the temperate parts of Europe. It attains the highest perfection in certain districts of the country, where summer heat seldom exceeds 75°. With good cultivation they grow to a large size, are very luscious, and much appreciated as dessert. A dozen of the largest and best flavoured varieties for exhibition are: (Red) Industry, Keen's Seedling, Rifleman, and Warrington; (White or Green) Bright Venus, Hedgehog, Snowdrop, and Whitesmith; (Yellow) Early Sulphur, Leader, Leveller, and Mount Pleasant. The Gooseberry thrives best in a free loamy soil on a clay subsoil in an open situation, and bears the best fruit on young well-ripened wood of the previous year. If they run to gross wood, the bushes should be lifted and transplanted, which brings them into fine bearing condition. The shoots should be thinned out in summer, and the only winter pruning will be the partial shortening of the fruit-bearing shoots. After the fruit is set, the ground should be mulched, and the bushes liberally fed with liquid manure while the crop is swelling. When the fruit begins to ripen, no water should be given, and some varieties require protection from heavy rain, to prevent bursting. The crop should be thinned if necessary, and the flavour is much improved by a full exposure to the sun while ripening. They make a fine dish when set up in conical form, with the stalks out of sight. The points of merit are: (1.) Flavour; (2.) size; (3.) colour, and thinness of skin.

THE GRAPE.

The successful cultivation of the Grape Vine is considered by many to be the masterpiece of horticultural skill; and although possessed of a strong constitution, and capable of adapting itself to many conditions of soil and circumstances, yet there is no plant which so soon shows the results of unskilful management, or so well repays the grower for liberal treatment and any special care bestowed on it, by the well-developed and highly-finished clusters of luscious fruit. There are several well-marked stages of development in the vine and its produce, from the time of starting to grow till the fruit is gathered, each requiring its special line of treatment and entailing a considerable amount of care, skill, and close observation to foresee results and keep clear of all that tends to interfere with perfection in the Grape. Being the most popular of all hothouse fruits, the culture of the Grape is entered into with great zest by those amateurs who can afford the luxury of a vinery, and by every professional horticulturist who has an opportunity. For exhibition purposes, however, its cultivation is carried on by a comparatively small proportion of the total number of growers, which might be greatly increased, to the mutual advantage of all concerned.

In selecting the best varieties of grapes for exhibition, the Muscat of Alexandria may well hold the premier place. Its large and handsome-shaped bunches and berries, beautiful colour, richly luscious flesh, and high flavour are not found combined to the same extent in any other variety. The Black Hamburgh keeps close company with the Muscat, slightly behind it in some points, while excelling it in others, and especially in general utility. The many excellent qualities of Madresfield Court claim for it a place next the Hamburgh among black grapes, while among white varieties Buckland's Sweetwater, in its best form, may be placed next the Muscat, although a long way behind that superior grape. Among other white grapes, Foster's Seedling, Golden Queen, Mrs. Pearson, and Tokay are good useful grapes, while Syrian, Trebbiano, and White Nice bear large handsome bunches, but being deficient in quality and flavour, they may be considered third-class grapes. Duke of Buccleuch is a fine early white grape, with splendid berries, but difficult to get in first-rate condition; and the Duchess of Buccleuch, although small in berry, is unrivalled for flavour. Among other black varieties, Lady Downes, Muscat Hamburgh, and Mrs. Pince, when in first-rate condition, come near the top of the list. Alicante and Alnwick Seedling bear handsome bunches of fine berries, with a dense bloom, while the enormous bunches of Gros Guillaume, and the enormous berries of Gros Colman and Gros Maroc, when well finished, carry them well to the front in competition, and from a commercial point of view the last two might almost top the list.

To grow grapes to the highest state of perfection for exhibition, the vines must be comparatively young and in vigorous health, with the borders fresh and well made. Worn-out vines and exhausted borders are totally incapable of producing first-class grapes, and must be renewed if the grower is to have any chance to command success. In choosing the site of a vinery, an open position with a gentle slope to the south is to be preferred. A lean-to house of the following dimensions makes a first-rate vinery for all ordinary purposes—Height at back, 15 feet; at front, 3 feet; with a width of 15 feet; giving a length of rod of vine of about 21 feet. This structure should be divided into convenient lengths, but not less than 30 feet, which is sufficient for growing three or four kinds; and more varieties should not be grown in the same house if it can be avoided. The earth should be excavated to a depth of about 3 feet 6 inches, and thus allow 3 inches for concrete in unsuitable subsoil for vine roots; 9 inches for perfect drainage, and 2 feet 6 inches for the soil in which the vines are to grow. Deeper borders are unnecessary, and in strong soils the depth may be reduced to 2 feet of soil with advantage. Surface dressings, liquid manure, and the beneficial influence of air and light operate most effectually on the roots in comparatively shallow borders, and lengthened experience has fully proved the good results. The concrete, where necessary, is composed of a layer of lime and gravel or ashes, well mixed in the proportion of about one of lime to six of gravel, laid on while the lime is hot, levelled and beaten smooth, and allowed a few days to dry and consolidate, when it forms a hard impervious bottom, and should have sufficient slope to run water off quickly. On this the drainage is carefully laid, from six to nine inches deep, according to the slope of the bottom and the nature of the materials. The top of the drainage fairly levelled, and a turf with the grass downwards laid all over it, makes all ready for the soil. This is usually obtained from an old pasture by stripping the top three inches or more, according to the quality, a moderately stiff marly loam being the most suitable. It should be chopped into pieces about four inches square, and mixed with a little lime and brick rubbish, according to its nature, to keep it open and allow water to percolate freely. A little wood-ashes may also be added, but as a rule no organic manures should be mixed with the soil when making the border. Too rich a soil has a tendency to cause over luxuriance, which is not conducive to fruitfulness. The proper aim should be to lay a good foundation on a moderate growth of well-ripened wood. The compost having been carefully put in its place and well consolidated, the border will be in readiness for the planting of the vines. Many varieties are often seen planted in the same house, no consideration having been given as to whether they were amenable to the same treatment or not. This error should be strictly avoided, and only those varieties planted in the same house which are known to succeed under the same treatment and temperature. One or two houses should be devoted to those requiring a high temperature, such as the Muscat of Alexandria, Mrs. Pince, Muscat Hamburgh, Gros Colman, Gros Guillaume, and Trebbiano. Although the three last are not high-flavoured grapes, they are improved by the same treatment as Muscats. Other houses may contain Alicante, Alnwick Seedling, Black Hamburgh, Buckland Sweetwater, Golden Queen, Madresfield Court, and their like.

The houses are, say, 30 feet long, taking nine vines, each 3 feet 4 inches apart. The roots of the vines should be carefully examined before planting, to ascertain if they are in a healthy state and likely to make a good start. If the canes are long, they should be bent horizontal, and the eyes cut out to within a foot of the roots. Early in March the vines should be placed in a gentle heat, and when they have grown about three inches they are ready for planting; shaking them carefully out, pruning any long roots, and spreading all carefully out, and covering them with some of the finest of the soil, settling it about them with tepid water, and finishing off all smoothly. After the roots have got hold of the soil, watering must be carefully attended to, as neglect of this is a frequent source of failure in vines, especially at first, when the border is extremely porous, and water percolates freely to the drainage. The strong roots in searching for water push their way downwards until they form a perfect mass at the bottom of the border, choking the drainage, and causing late starting, late ripening, and inferior fruit, with more or less shrivelling.

It is found that by lifting the roots of unsatisfactory Muscats and laying them near the surface, the vigour of the plant and the quality of the fruit is restored to its normal condition. This may apply with more or less effect to other grapes, but it is especially noticeable with Muscat of Alexandria. To prevent the roots going down, the border should be mulched and frequently watered. Close attention must also be given to the atmospheric conditions. A temperature of 50° at night and 60° by day should not be exceeded until the leading shoot is about two feet long, and the roots taking kindly to the soil. It is of the utmost importance not to exhaust the sap in the stem by too much heat until the roots are in a condition to supply the demand. Otherwise the plants may survive, but a check is given to them from which they will not soon recover. When the roots are actively at work, 5° more of heat may be given both day and night, with plenty of moisture kept up by syringing in the morning and afternoon, and damping the border several times daily. By the time the shoots are half way up the roof, the temperature should be 60° at night, 70° to 80° with sunshine during the day. All laterals should be pinched at the second leaf, and the sub-laterals at the first leaf, which is much better for the vines than the detrimental practice of allowing the laterals to grow without stopping until the roof is entirely covered. The sap should be concentrated as much as possible in the stem, to be utilised in producing the growth of the following year.

Due regard having been paid to watering, airing, temperature, and keeping insects in check during the summer, the shoots should be well ripened and ready for pruning by December. A perfect rest of at least two months without heat should be given to all vines, whether old or young. In pruning, an eye should be selected on each side of the rod near the bottom of the rafter, from which to form the future spurs, and an eye above to form the leading shoot. All loose bark should be removed, and the rods washed and dressed with some approved insecticide. Forcing should commence about the middle of

February, with a temperature of 55° at night, 60° to 70° by day, increasing to 60° at night as the leaves expand. By midsummer the heat should rise to 65° at night and 80° to 90° with sunshine during the day for the Muscat house, and the cooler houses always kept 5° lower. All laterals should be regularly pinched, and the side-shoots to form the spurs should be stopped at the sixth leaf and tied to the wires as soon as they can be safely handled, but great care must be taken, as they snap off very easily. A few of the shoots lower down the rod should be left, and pinched at the third leaf so as to clothe the stem, but all bunches should be removed. The treatment of the vines in the second year is exactly similar to that of the first year.

The third or fruiting year may be regarded as the most important, as then the whole circle of treatment is carried out. The side-shoots should be pruned at the first eye, or second if most prominent, and the leading shoot shortened to about 2½ feet, which will furnish two laterals on each side about 14 inches apart. Muscats and other white kinds succeed best, however, with the spurs 18 inches apart, to admit plenty of light to assist the colouring of the fruit. This is also necessary to the perfect colouring of some of the black kinds with thick tough skins.

The advantages gained by moderately close pruning are: (1.) Securing a gradual thickness and strength from the root upwards; (2.) ensuring a regular break; (3.) admitting more light and a freer circulation of air among the vines; and (4.) as a prevention of overcropping, by restricting the number of bearing shoots while the vines are young. After the rods are cleaned and dressed, the surface soil should be taken off the inside border down to the roots. The border may have subsided, and should be made up with turfy loam mixed with half-a-peck of artificial vine manure and the same quantity of bone-meal to the barrow-load of soil, and with some decomposed manure bring it up to the original level, finishing off with a thin layer of fine soil and wood-ashes.

In order to obtain perfectly finished grapes in August, forcing should commence in the Muscat house towards the end of January, with a temperature of 60° to 65°, and in the second house in the middle of February, with 5° less heat, rising gradually with the advance of the season. Some flowers will be opening by the middle of April, when the heat should range from 65° to 70° at night, with a rise of 10° during the day and a drier atmosphere. This is an important stage in the growth of the grape. On the successful fertilisation of the flower depends the symmetry of the bunch and size of the berry, two properties which contribute largely to the formation of a perfect bunch. A successful set is believed to be largely due to the favourable state of the weather at the time of flowering; but experience teaches us that the use of the camel's-hair brush is a great assistance in effecting this desirable result, by removing any exudation from the stigma, and diffusing the pollen equally over the flowers. The transference of the pollen of one variety to another, and from those with plenty to others that are deficient, is beneficial in perfecting fertilisation. Previous to flowering, remove all bunches except the best one on each lateral, and allow none on the leading shoot. After setting, a further reduction of the bunches is necessary to obtain success and to do justice to the vines. It is not usual to lay down rules for the cropping of vines, so much depends on the strength of the plants and other circumstances; but as we are dealing only with those which are in the very best condition, we may endeavour to define what should be a fair crop on vines three years old or the first season of bearing. Although the perfect finish and colour of the fruit may be a good measure of the capacity of a vine, it may be all that could be desired, and yet the vine be so much overcropped as to very much impair its vitality. To produce exhibition bunches, all vines of equal strength should carry the same weight of grapes, say from six to seven pounds, at three years old. The berries should be thinned as early as possible after they are set. This operation is not easily described, but a few observations will be of some service to those not thoroughly conversant with the process. The object of thinning is to obtain shapely bunches and large berries, giving just as much room to each berry as it requires for attaining its full size.· If it is possible to complete the operation at once, it should be done; but to secure equal berries it is often necessary to leave more than is required in view of a second thinning. This should be effected as soon as it is clearly seen which berries are likely to attain a full size. Large berries on strong foot-stalks are not readily over-thinned. Long lopping shoulders should be cut in, and may be slung if necessary to keep them square; but tying up generally should be avoided, as loose bunches are of less value in competition than those of a firm and compact build. When the berries are swelling freely, a temperature of 65° to 70° at night, with 75° to 80° during the day, should be maintained, while 90° may be reached with strong sunshine and plenty of ventilation. At this stage, and previous to colouring, surface dressings of different kinds of approved manures should be given. Besides manures specially prepared for vines, fish manure, bone-meal, and Peruvian guano all contain elements of great value in the successful cultivation of the vine. These should be applied thinly but often to the surface of the border, and watered in. By the time the berries are the size of peas, it will be necessary to maintain a night temperature of 70°, and during the day 80° to 90° in sunshine. Muscats and some other varieties are liable to have their leaves scorched about this time, especially after dull days. Free ventilation night and day, and a lower temperature and drier atmosphere for a few days, will prevent scorching of either leaves or fruit.

The airing of vineries is, as a rule, not sufficiently studied, and the temperature is too often allowed to rise to a certain degree with the sun shining before the ventilators are moved at all. This is a bad practice, considering the state of the atmosphere in houses of modern construction, and is a prolific source of scorching. When vines are in full growth, top ventilation should be given as soon as the thermometer indicates a rise from the heat of the sun, but not in excess to cause the mercury to fall. The temperature may then rise 5° before admitting front air or giving more at the top. In order to keep the atmosphere sweet and moist, water should be sprinkled on the border and path several times daily.

Shutting up early in the afternoon is recommended by some to utilise the natural heat and save the artificial. This, in a certain measure, is to be commended; but in closing the ventilators while the sun's rays are so strong as to raise the temperature to 100° or more, and then deluging the house with water through the engine or syringe, amounts to what is termed the "steaming system," and is not at all a wise or natural process. This saturated atmosphere may be congenial to rapid growth, but when it is made at the expense of the perfect elaboration of the sap and consolidation of the tissue, a check is given to the functional operations, causing general derangement, from which can be traced some of the disorders of the vine which are often attributed to other causes, and consequently excessive moisture in a close atmosphere should be avoided. During the stoning period, undue forcing is certain to produce scalding in the berries of varieties liable to it, but a free circulation of air both night and day with 5° less of heat will prevent its occurrence to any great extent. When the berries are stoned and begin to swell, the former conditions as to heat, moisture, and manuring should be resumed. When signs of colour appear, a copious watering should be given to any varieties of which the berries are liable to crack. After this they should be kept comparatively dry till the ripening is complete, when sufficient water may be given to keep the roots active and healthy without any danger of the berries cracking; but to prevent injury by damp, the watering should be done on a dry clear morning. Other varieties may be watered when necessary throughout the ripening process, and some dry mushroom-bed manure spread on the border to prevent evaporation. Keep up a free circulation of dry warm air at night by leaving a small opening at top and front, keeping up the same temperature of 70° for Muscats and 65° for the others, to be gradually reduced as the final stage is reached.

A careful watch should always be kept for the appearance of red spider, which should be washed off with a sponge and soapy water to prevent it spreading. Syringing with insecticides is now unsuitable, and even rain-water, from having been in contact with houses and dirty tanks, cannot be used without leaving marks on the berries. The safe course is to examine the leaves carefully previous to the fruit colouring. If they are quite clear of red spider, little danger may be apprehended from it afterwards. Thrips are also injurious to the vine. They attack and weaken the leaves and disfigure the berries, but are far more easily removed than spider. Fumigating two or three times on alternate evenings will suffice to kill or keep them in check.

All the light possible should be admitted to Muscats and thick-skinned grapes when they are colouring, by putting the leaves aside, taking care that the direct rays of the mid-day sun do not strike on the berries, as it is apt to discolour them. The condensation of moisture on the berries must also be prevented by keeping the pipes moderately warm at night and increasing the ventilation in the morning before the temperature rises. Although the atmosphere of the house may appear to be perfectly dry, there is always more or less moisture in it, which heats sooner than the berries, and a deposition of dew immediately takes place on the cooler surface, which soon evaporates and leaves a mark on the bloom.

Beside the effects of the pests already enumerated, grapes are liable to suffer from attacks of mildew, rust, scalding, and other complaints, which disfigure their appearance and reduce their competitive value.

Mildew is an insidious disease, and its presence is sometimes not easily accounted for. Certain atmospheric influences accelerate its progress, such as a damp, close, sultry air, while even extreme dryness favours its spread. Its appearance on the leaves resembles specks of pale grey powder, and it is sometimes found on the berries when the leaves are healthy and clean. When first observed, flowers of sulphur should at once be applied with an elastic puff, and if the berries are kept dry, the sulphur can be blown clean off them.

Rust is produced by injury to the cuticle of the berry whilst young and tender. Its appearance is attributed to various causes, such as the application of sulphur to hot-water pipes, the handling of the berries while thinning, a cold current of air, and an over-moist atmosphere; but whether any of these be a cause of it or not, they should be all carefully watched, and as much as possible avoided.

Scalding is an injury to the berry at a more advanced stage of its growth. The main cause of it is an unnatural condition of the air through the want of a free circulation. Too high a temperature in the morning before ventilation is given, and too early closing with much heat in the afternoon, are fertile sources of it, and must be strictly guarded against.

Shanking is a disease affecting the foot-stalks of the berry. It appears in two forms and at different periods; first about the time of stoning, when a dark line may be observed moving along the centre of the stem or foot-stalk in the direction of the fruit, soon after which the berry begins to shrivel. The second begins about the time of colouring by a dark ring encircling the foot-stalk, which ultimately turns black throughout. Instead of the berries shrivelling, as in the other case, they retain their plumpness for a considerable time, although quite useless. Muscat of Alexandria is most subject to the first form of disease, while the second attacks the Black Hamburgh and other varieties. Defective root-action is probably the origin of the first, and the second may be ascribed generally to a want of vital energy in the plant, possibly from lack of nourishment, or the weight of the crop being greater than the plant can sustain. In well-managed vineries this disease rarely occurs.

Shrivelling is a natural condition of all grapes if allowed to hang long enough, but when it appears prematurely it is obvious something is wrong. Early shrinking of the berries chiefly arises from the imperfect condition of the saccharine matter, resulting probably from the adverse circumstances under which the fluids have been assimilated in the leaves, and thence secreted in the berries. A dry warm atmosphere, or any check to leaf and root action, is productive of shrivelling.

It is imperative that every precaution be taken to preserve the grapes from the attacks of insects and all diseases which blemish and spoil their appearance, so that they may be staged without a speck and in the highest state of perfection.

We have explained briefly the principal operations and matters of importance to be observed in the cultivation of the vine during the first three years of its growth, and refer the inquirer after fuller details to Mr. A. F. Barron's "Vines and Vine Culture," the best treatise extant on the subject.

A few hints on preparing the grapes for exhibition may be useful to the beginner. The first thing to determine is the stand on which the grapes are to be exhibited, and Horticultural Societies should insist on these being of a uniform slope and dimensions. The irregularity and untidy appearance of so many different forms and sizes as are frequently seen at exhibitions detracts greatly from the pleasing effect of the display, and even a good bunch set on a badly constructed stand is shown at a great disadvantage. A convenient size, with a slope to show off the fruit effectively, should be adopted at all exhibitions. The following are good average dimensions :—Height at back of sloping board, 12 inches ; breadth of slope, 16 inches ; and length of stand 9 inches for each bunch it has to carry, but the length may vary according to the size of the bunches. The back or upright board should rise $2\frac{1}{2}$ inches above the junction with the sloping board, making it $14\frac{1}{2}$ inches high. The ends should be filled in with brackets to keep the whole firm. The top of ,the stand should be covered with plain white paper neatly pasted down, but with no ribbons or brass nails used in the way of adornment, which are out of place and unnecessary.

The bunches should be cut with a piece of the shoot attached, and great care must be exercised in handling them, so as to preserve the bloom without the slightest rub or injury. Examine the bunch closely, and place the best side uppermost, fastening it securely at the top, and also near the bottom with narrow tape carefully passed between the berries and through two small holes in the board, to be firmly tied behind. Neat handling by the operator is necessary to avoid rubbing the berries, and to have the bunch held perfectly firm and secure on the board, so that it cannot roll in the least during the joggling of a long journey by road or rail. After the bunches are securely fixed on the stands they should be placed in a strongly constructed box, specially prepared for their safe transit to the exhibition, and in which they can be set expeditiously and securely. A good size for such a box, which two men can easily carry, and to hold eight bunches on their stands in two rows, with the back of the stands to the sides, is 36 inches long, 15 inches deep, and about 30 inches wide, all inside measures. It should be fitted inside with slips and grooves, between which to set and fasten the stands quickly and securely. The box should be provided with strong handles for carrying it, and with a proper lock and duplicate keys.

The points of merit in grapes are: (1.) Size and form of bunch; (2.) flavour, size, and quality of the berries; and (3.) colour and bloom. A model bunch of Muscat of Alexandria may be described as follows:—Length, from 12 to 15 inches; breadth at widest part of shoulders, two-thirds of the length, and tapering regularly to the point; berries large, even in size, not squeezed, but so closely set that the bunch retains its perfect shape when laid on the stand; of a clear amber colour, with a perfect bloom, and free from all blemishes. This description applies generally to all other grapes, making due allowance for their natural variations.

THE MELON.

The Melon is said to be a native of the moist parts of the low valleys lying to the south of the Caucasus and the Caspian Sea in Asia, where it flourishes with great luxuriance. From thence it has spread through cultivation to all parts of the world with a tropical or semi-tropical climate. Wherever there is sufficient summer heat to bring the Melon to maturity it thrives well and is cultivated with success, even although the winters may be far more rigorous than in Britain, where, to be successful, it has to be grown in a hothouse. It has been cultivated for its fruit in this country since about 1570, but it is only within the present century that it has reached its highest state of perfection, and become so popular among us as a dessert fruit. The demand for it is now so large that it far exceeds the supply raised at home, and it is imported in great quantities from warmer countries, where it is cheaply grown in the open air. Like the Pine-apple, however, the Melons grown in our hothouses much excel the imported fruit in richness and high quality.

As an exhibition fruit the Melon is highly esteemed, and a well-grown specimen, in perfect condition, is a strong point in a collection. It forms a handsome dish, and is a good counterpart to the Pine-apple. Almost every competitor has a favourite variety, which he cultivates with much diligence and success; and so much depends on the careful attention paid to the saving of the seed, that a grower for competition is, as a rule, better to save his own. Among the popular varieties which have acquired a reputation for the exhibitor's purpose are the following :—

Green-fleshed Varieties.—Eastnor Castle, La Favorite, Longleat Perfection, and William Tillery.

Scarlet-fleshed.—Benham Beauty, Hero of Bath, Read's Scarlet, and Sion House.

White-fleshed.—Colston Bassett, Hero of Lockinge, Lord Strathmore's Favourite, and The Countess.

For the cultivation of the Melon to the highest state of perfection in this country, it must be grown in a glass-house efficiently heated by hot water. The best form of house is a low lean-to or a half-span, facing south, so that it may receive all the light and sunshine possible in our variable climate, as without these two important elements the Melon is seldom superior to a common gourd. The heating apparatus must be able to supply an abundance of top and bottom heat, as may be required by the Melons in all conditions of the weather. A brick bed about 2 feet wide and 18 inches deep, well-drained, and with a flow and return bottom-heat pipes beneath it, should be formed along the front of the house, and as near to the roof as is convenient for attending to and training the plants to the wires, which should be fixed about 15 inches from the glass, about 10 inches apart, and kept tight by screwed bolts and nuts.

The Melon delights in a strong mellow loam, of a calcareous nature, and enriched with decomposed vegetable matter. A compost of strong fibry loam with a third of well-decomposed leaf-mould and a sprinkling of old lime rubbish, all well mixed and packed rather firm in the bed, grows it well. For early exhibitions, in May or June, the seed must be sown in January and February in the same compost, putting a seed in a three-inch pot, or planting them two inches apart in pans, covering the seed with a pinch of sharp sand, and setting the pots or pans in strong bottom-heat till the seeds vegetate. As soon as they are fairly above ground they should be placed close to the glass to keep them sturdy, and care must be taken to shade them from the sun. When they have made the first leaf, they should be re-potted or potted singly into five-inch pots, using the compost in a warm mellow state, and being careful not to bruise the stem of the plant while firming the soil. Every precaution must be taken not to bend or bruise the stem when it is young and tender, as it renders the plant more liable to disease or canker. Grow then in plenty of heat near the glass, and give air on all favourable occasions to keep the plants healthy, vigorous, and sturdy. The plants will be ready for planting out in the bed in about six weeks from the time when they were sown. The bed should be made up a few days beforehand, to warm the soil to the desired temperature, about 75°. In planting, set the plants about two feet apart, and place the earth firm around the balls, which should not be covered, but just level with the surface of the bed. Carefully train the plant up the wires till it has reached within a foot of the top, when it should be stopped, and the fruit-bearing laterals will soon push forth along its entire length. The laterals should be all stopped at the second or third leaf; and as soon as the female flowers expand, they must be gone over daily about one o'clock, and "set" or impregnated, by picking off a male flower charged with pollen and placing it on the stigma of the female flower or fruit. As soon as four fruits are set on a plant and swelling freely, no more need be set, and any more swelling should be removed. For exhibition specimens, two to four fruits on a plant are as many as it can bring to perfection. From the beginning water must be carefully applied, and a moist temperature kept up of

M

about 70° at night, rising to 80° during the day, and to 90°
with sun-heat, and plenty of air; shading when necessary to
prevent flagging and scorching, but giving all the light the
plants can bear, especially as the crop nears maturity. If
the plants are maintained in a healthy, vigorous state, insects
are not troublesome, but if red spider appear, it should be at
once attacked with the syringe or the leaves carefully sponged,
so as to keep it in check. As the fruit begins to ripen, the
water must be gradually withheld, and all the light and air
given that is possible while the high temperature is maintained, and by this means a rich flavour and fine quality is
secured. Melons are quite amenable to the stimulating influences of liquid manure, but it must be cautiously applied, so
as not to induce rank growth, deteriorating the quality of the
fruit, and often ending in disease. Before the fruit is half
swelled, each should be supported on a piece of small-meshed
galvanised wire-netting about six inches square, and slung by
strings from the trellis at such a distance as the fruit may lie
on its side and have perfect freedom to swell to its full size.
Cracking of the fruit is the result of careless watering, and
under proper management need not be apprehended. Should
the fruit ripen too early for the show, it may be cut as soon
as it reaches maturity, but not over-ripe, and hung up in a
net in a dry cool room, where it will keep sound and good for
a fortnight or more with scarcely any loss of flavour. To
have a Melon at its best for flavour on the day of the show,
it should be quite ripe and cut two or three days previously,
and placed in the warmest spot in the house in the full blaze
of the sun till it is wanted. Later in the season, for exhibiting at shows held from July to September, the Melon requires
exactly the same treatment, but it takes less time in summer
to reach maturity, about three months sufficing from the time
the seed is sown until, under good management, the fruit is
ripe and fit for exhibition.

The chief points in a Melon are : (1.) Shape, which should
be handsome, smooth, and globular, and the netted varieties
regularly marked ; (2.) size, which should not be less than
18 inches in circumference ; and (3.) the flesh thick, tender,

sweet, juicy, and melting, with a rich, luscious flavour, a pleasing aroma, and a thin skin.

THE NECTARINE.

This is a very distinct and beautifully coloured section of the Peach tribe, *Amygdalus persica*, the piquantly refreshing, rich juicy flesh of which is preferred by some palates to that of the more luscious typical section. As an exhibition fruit, the Nectarine ranks next after the Peach. For that purpose the following are among the best, arranged in the order in which they usually ripen :—Lord Napier, Elruge, Pitmaston Orange, Pine-Apple, Humboldt, and Spencer. The cultivation of the Nectarine is the same in all respects as that of the Peach, for details of which see the article on that fruit.

The points of merit in a dish of Nectarines are : (1.) The fruits should be large, equal in size, smooth, and symmetrical; (2.) ripe and sound, with a rich juicy flesh of a brisk vinous flavour, and parting freely from the stone, which should be small ; and (3.) the colour should be bright and fresh, dark red next the sun, and orange or yellowish green in the shade.

THE ORANGE.

There are several species of the genus *Citrus* cultivated as oranges, the origin of which is lost in antiquity ; but from the earliest ages of which there is any record, they appear to have been grown and used as food by the natives of Eastern countries from Asia Minor to Japan. At the present time it it is extensively cultivated for commercial purposes in the southern countries of Europe, the Madeira Islands, Florida, and Southern California, besides many other places of less fame as orange-growing centres. Its cultivation in this country as a dessert fruit has never attained any amount of popularity, or received anything like the attention from horticulturists which its excellence under proper treatment deserves. To any one who has tasted the delicious richness of flavour of a perfectly ripe and full-grown orange, fresh gathered from the tree, grown in a suitable hothouse, it is a marvel in those days of cheap glass and advanced cultural skill why the Orange is not cultivated with success in the gardens of the wealthy, where it might well fill a place among less delicious fruits, which are often cultivated at much more cost than it takes to grow fine crops of first-rate oranges.

As an exhibition fruit, the ordinary produce of our hothouses has seldom been such as to recommend it for a high award; but with suitable accommodation and proper treatment, specimens can be easily grown to rank alongside of the best peaches, figs, or melons, if not to equal the grape itself. The best varieties for the purpose, so far at least as the numerous fine kinds grown abroad have been tested by cultivation in Britain, are the *Maltese Blood*, *Mandarin*, *St. Michael's*, and *Tangerine* Oranges. Where these varieties can be grown in a properly constructed orangery—very unlike the old-fashioned dungeon called an " Orangery "—they amply repay with abundance of luscious fruit the cost and skill bestowed on them. They thrive to perfection in a rich marly turfy loam, in which apples grow with the greatest success. In preference, the trees should be planted in a well-drained border, made of such soil, with the addition of a free sprinkling of half-inch bones. In their treatment, they require much the same temperature and management as the peach when forced; but being evergreens, even more care is necessary to guard against the inroads of noxious insects. which are their greatest pest, and difficult to eradicate when once they get a footing among the evergreen foliage and branches. With that exception, they are much easier managed than most forced fruits, and require nothing but average attention to heat, air, moisture, thinning of fruit, watering. and manuring. When they are swelling a crop of fruit, they are much benefited by a judicious supply of clear liquid manure. They require very little pruning or training, merely sufficient to keep them in shape, with the branches wide enough apart to allow a free play of light and air around the ripening fruit and the blossom of the succeeding crop, which are usually both upon the tree at the same time. The culture of the Orange in pots is exactly similar, and with liberal treatment they produce fine crops of beautiful fruit. The Orange thrives best when not over-potted; but the drainage must be ample, to allow of copious supplies of liquid manure, as well as of clear water, at the seasons when they are required. The plants must never be allowed to become dry, but careful watering is necessary in winter to avoid souring the soil. At starting in spring, a mild bottom-heat to the plants in pots is beneficial. Their points of merit are: (1.) Size; (2.) flavour and juiciness; and (3.) colour and thinness of skin.

THE PEACH.

This delicious fruit has been cultivated in Oriental countries from time immemorial, and was generally considered to be a native of Persia; but De Candolle, on what appears to be irrefragable evidence, traces its origin to China, where it is mentioned by Confucius as early as the fifth century before the Christian era. It is now cultivated with success in most of the warmer parts of the temperate zones, and especially in the United States of America, in some parts of which it forms a staple industry, and grows with great luxuriance and fruitfulness under the rudest methods of cultivation. In Britain, however, it must be grown under glass with proper heating appliances, so as to command the greatest success and certainty of a crop. In some favoured spots it may be grown with tolerable success on open walls, where it generally attains a high flavour, although usually lacking some important point deemed necessary in first-rate exhibition fruit.

The Peach and Nectarine are merely varieties of the same plant, *Amygdalus persica*, but they differ so markedly in their appearance and flavour that separate classes are always assigned to them at horticultural exhibitions; therefore they are treated separately here, although their culture and management are precisely similar. Among the best varieties for exhibition are Hale's Early, Royal George, Stirling Castle, Dymond, Noblesse, and Sea Eagle. A similar house to that recommended for the vine, with heating power according to what is wanted, grows peaches to perfection. If the trees are well established in good soil, no preparation of a special nature is necessary, as well-managed old trees in vigorous health will produce quite as fine fruit as the best of young trees. Should it be necessary, however, to plant young trees in a fresh border, care should be taken to select the best varieties worked on proper stocks, to avoid failure in bearing by some varieties, and they should be properly planted in a border made of a sound, medium, turfy loam of a calcareous nature, taking care to have the border well drained. If the loam is rather light, it will be improved by an addition of marl or clay, and if deficient in calcareous matter, lime rubbish or similar material may be added with advantage. Rich borders, causing rank growth, are quite unsuitable, and ought to be avoided if the greatest success is desired.

For the earlier shows in May and June it is necessary to start the Peaches at the beginning of the year, later houses taking up the succession for July and August, and cool houses furnishing Peaches for exhibitions in the autumn. The earliest house should be started in the first week of January to secure Peaches by the middle or end of May; and at first the night temperature should not exceed 45°, rising to 60° at mid-day, with plenty of air on. The trees should be well syringed morning and evening, and a moist healthy growing atmosphere maintained. As the trees come into flower the syringing should be suspended, and a drier air maintained till the crop is set; but a parching heat must be avoided, as it will certainly bring on an attack of red spider, one of the worst foes to the culture of the Peach in hothouses. A daily use of the camel's-hair brush, about one o'clock when the flowers are dry, helps much to secure a free set of early peaches.

To assist the trees in setting a good crop of fruit, strip the flower-buds from the under-side of the shoots, and keep up a free circulation of dry air when the flowers are expanded. When the crop is well set, disbudding should be gradually performed, to prevent crowding and to secure bearing wood for the following year. Remove all shoots not required, except those starting from the base of the fruit, which should be pinched. Thin the fruit when it attains the size of peas to about double the number to be eventually left to ripen, which should be in the best position for the sun to colour and flavour them properly. The cropping of peaches by calculating so many to the square yard is not a good method to recommend when the circumstances are so varied and strength of trees so difficult to define. Young vigorous trees are the better to be rather heavily cropped, while old trees of moderate growth would be injured by a similar number per square yard. It should be remembered that one good-sized peach is better for any purpose than two small ones, but more especially for exhibition. After the final thinning has been completed, the roots should be watered if dry, and a good mulching of rich manure given to assist in feeding and keeping the surface soil sweet and moist. Syringing should be done in the morning, and early enough in the afternoon to allow the foliage to get dry before night; and if aphides are troublesome, the ordinary solution of soap and paraffin should be applied. When the fruit begins to swell, the roots should be supplied with liquid manure from drainings of the farmyard, and some of the best artificial manures for fruit-trees. This should be carried on as it is required till the approach of maturity, when syringing should cease, and water be withheld as far as possible with safety to the crop and trees.

The cultivation of Peaches and Nectarines on the open wall is exactly similar, although trees on walls are not generally so well attended to as those in houses. The crops on open walls are very precarious, and cannot be relied upon, even under the most favourable circumstances of season and climate. Early varieties are more easily cultivated with success than late ones. They should be planted on a south wall, kept well thinned out, and the young shoots trained close to the wall as they grow. The thorough ripening of the wood and the protection of the blossom in spring are the chief factors to success in the culture of the Peach and Nectarine in the open air.

The fruit should be carefully handled in gathering it, because undue pressure on any part causes a blemish to appear in a very short time. Shallow boxes to contain a single layer should be used for packing them for exhibition, and each fruit, wrapped first in tissue-paper and then in a small square of cotton-wool, and placed side by side among soft paper shavings, or any other suitable material, covering and making all secure with the same.

After the fruit is gathered, and to secure a good crop the following year, the shoots not required to bear fruit should be cut clean out, and all laterals from the current year's growth cut back to one eye. Syringe the trees to keep them clear of insects, water the roots, and attend to the ventilation in order to secure perfectly ripened wood.

The points of merit are : (1.) Size ; (2.) luscious flavour ; and (3.) colour.

THE PEAR.

The Pear is a native of Britain and other parts of Europe and Asia within the temperate zone. It is said to have been widely cultivated by the Greeks and Romans more than three thousand years ago. The warm districts of France seem to suit the finer varieties to perfection, while the more hardy kinds produce splendid crops of fine fruit in many districts of Scotland. It is also one of the most accommodating of fruits, as by a careful selection of varieties an excellent supply can be provided during two-thirds of the year.

Among the best pears grown without the aid of glass for exhibiting in a ripe state in the autumn are Beurre d'Amanlis, Fondante d'Automne, Jargonelle, Madame Treyve, Souvenir du Congrès, and Williams' Bon Chrétien ; and of the later dessert varieties to be exhibited, "ripe or unripe," the following are excellent : — Bergamotte Esperen, Beurre d'Aremberg, B. Bachelier, B. Bosc, B. Clairgeau, B. Diel, B. Rance, and B. Superfin, Doyenne du Comice, Duchesse d'Angôulême, Easter Beurre, Emile d'Heyst, Glou Morceau, Hacon's Incomparable, Josephine de Malines, Louise Bonne of Jersey, Marie Louise, Ne Plus Meuris, Olivier de Serres, Passe Colmar, Pitmaston Duchesse, Thompson's, and Winter Nelis. The best culinary pears for exhibition are Bellissime d'Hiver, Black Worcester, Catillac, Uvedale's St. Germain, Verulam, and Vicar of Winkfield. The finest variety for competition in each of these three sections, taking all points into consideration, is probably Jargonelle, Marie Louise, and Catillac.

A deep loamy soil suits the Pear; therefore, if the natural soil is unsuitable, it should be taken out, two to three feet deep and four feet from the wall, filling up with turfy loam of a medium texture. The trees should be planted in October, four feet apart, to be trained as upright cordons; and if the soil be dry, the roots should be watered and mulched. At the winter pruning the two strongest shoots should be reserved, cut back to 18 inches in length, and trained horizontally, leaving one bud six inches from the base, and another at the end of each shoot. These in time will form the four leading shoots, to be trained upright about a foot apart. The advantages from this method are, many varieties in small space, furnishing a large quantity of fruit in a short time, and the restriction of the roots enables the cultivator to water and nourish the trees more easily when it is required. By carefully pinching the young shoots in summer when two or three inches long, many spurs and fruit-buds will be formed, which in a few years will bear a heavy crop of fruit. When the fruit is swelling, it should be fully exposed to the sun, and the roots receive a liberal supply of liquid manure, and be top-dressed to bring the fruit to perfection.

The cultivation of many of the late varieties of pears, except during exceptionally fine seasons, is not attended with great success in cold districts: therefore an orchard-house, or any suitable glass structure, should be provided for them. They can be grown to a fine size in pots in an orchard-house, vinery, or peach-house, if not too much shaded by the overhead foliage. In order to have Beurre Diel, Marie Louise, Pitmaston Duchesse, or any of the late varieties ripe for autumn shows, they must be grown in heat. A vinery or early peach-house is suitable for their culture when grown in pots. Bush-trees of a suitable size, on the Quince stock, should be lifted from the open ground in early spring, root-pruned, and potted in 12-inch pots, using sound turfy loam and crushed bones. Place them in a light airy position in a peach or orchard house, and after the growth has been made, put them outside against a south wall, where they will get thoroughly ripened. During winter set them in a cool house, and start them with the first vinery in January. When they come into flower, they should be removed to a more airy house to set, as they will not do so readily in a moist atmosphere. When set, place them again in the vinery where the full light reaches them, and top-dress with rich soil, bone-meal, or any good artificial manure. When the fruit is growing freely, weak liquid manure should be given at every watering. By this treatment pears of a very large size and fine flavour can be grown and ripened in the same temperature as the vines.

Old-established trees on high walls, by careful management, may be put into condition to bear fine exhibition fruit, although they have for years been neglected. If the spurs be long, they should be gradually cut close in, extending the operation over three or four years, when some of the new spurs will begin to bear. Or by cutting out some of the old branches every year, and leading in young shoots, the whole tree can be renewed in a few years. Then by removing some of the soil above the roots, and working in fresh turf and bone-meal, and by mulching and feeding, excellent crops of fine fruit will result. The points of merit are : (1.) Size ; (2.) flavour ; (3.) shape. In culinary varieties : (1.) Size ; (2.) quality.

THE PINE-APPLE.

This is the handsomest as well as the most delicious of all the fruits produced in the greatest perfection, by the skill of the horticulturist, with the aid of modern appliances. It stands at the head of exhibition fruits, a position which its noble appearance, luscious mellow flavour, and rich perfume justly merits, and to the competitor it is an invaluable acquisition in a collection of the best dessert fruits. Although its cultivation has rather declined than increased during the past quarter of a century, it is still considered indispensable in all well-ordered garden establishments. The erroneous and altogether costly methods of culture formerly in vogue, and even yet not quite extinct, account in a great measure for the falling off in the numbers grown of this princely, wholesome, and useful fruit. When a more rational system of culture is generally adopted, and the cost of production reduced thereby to the same amount, or less, than that of other hothouse forced fruits, it will again become popular, and appear in every private garden with any pretension to efficiency and completeness.

The Pine-Apple is stated to be a native of the tropical parts of South America, but it has been so long introduced to the East and West Indies that it may be considered quite naturalised in them, and thrives in both with great luxuriance and fertility. In fact, it has now come to be looked upon as one of the staple fruit products of the West Indies, and, along with the Banana, forms the principal item in their rapidly increasing fruit export trade, from which much prosperity is anticipated in the future of those fertile islands. It was first introduced to Britain about 1690, but it was not till the year 1712 that we hear of it being grown for its fruit, about which time it was successfully fruited by Henry Telende, gardener to Sir Matthew Decker, at Richmond, Surrey. About twenty years earlier, Mr. Rose, the royal gardener, is represented in a well-known picture in the act of presenting a Pine-Apple to Charles II. as the first grown in England; but no other record has been found as to how, when, or where it was produced. In the "British Gardener's Dictionary," 1744, it is stated by James Justice that he erected in 1732 a pine-stove at Crichton, near Dalkeith, Midlothian, in which the Pine-Apple was fruited for the first time in Scotland. From that period to the present time it has been cultivated with the greatest skill and assiduity, and, with the help of improved appliances, it has been brought to such a high state of perfection as to rival the very finest produce of tropical countries, and to excel it in the important points of size and flavour in the finest specimens and best kinds.

Most kinds of Pine-Apples are, when well grown, suitable for exhibition, but at the present time exhibitors confine their selection to a few popular varieties. Among these, the Smooth Cayenne holds the leading place as an exhibition fruit, chiefly because of the ease and certainty with which it can be produced at all seasons of the year, and that, added to its fine size, handsome appearance, and excellent quality, makes it a favourite with all growers for competition. Next in popular favour is the Queen, which, when well grown and in its best condition, is a first-class fruit of the finest flavour and quality. After these a number of varieties are held in much the same esteem by competitors. For quality and richness of flavour, Black Jamaica still stands unrivalled; but as it seldom exceeds three pounds in weight in well-grown fruits, with neat small crowns, it is liable to be passed as undersized when exhibited in competition with large and well-finished specimens of other kinds of fair quality. One of the best of those latter is Charlotte Rothschild, a fruit very similar in all respects to the Smooth Cayenne, except that it has spines on the edges of its leaves, and on their merits the Smooth Cayenne is the most useful. Among others seen most frequently at exhibitions are Black Prince and Lord Carrington, two easily grown and fertile varieties, the first being a very tall conical fruit, requiring a high temperature and bright sunshine to ripen it all at the same time. Lord Carrington is also a conical fruit, and when at its best it makes a handsome specimen, although neither it nor Black Prince are to be compared in fine qualities to the four previously named varieties. When size alone is the desired object, a fully developed Providence Pine generally beats all competitors, and for this purpose it must be grown with the greatest possible vigour on a thoroughly matured plant to produce a fruit of the largest size and finest build. It takes two years or more to grow a proper plant that can be relied upon to give a fine fruit, and it must be carefully tended throughout to ensure success. Beyond its gigantic proportions the Providence has very little to recommend it, even at its best.

In the culture of the Pine-Apple a comparatively small amount of labour and attention is required, so long as they are given at the proper time, and all appliances are suitable and adequate. A hip-roofed house makes the best pine-stove, so constructed that the plants are close to the glass when full grown. The heating appliances must be ample for the work, to supply with ease the top and bottom heat in all states of the weather. The Pine is now generally grown in pots, and for all practical purposes it is undoubtedly the best system when properly carried out; but some very successful growers in bygone times preferred to grow them planted out in beds specially made up for the purpose, believing that they obtained larger fruit than from plants grown in pots. With more experience, there is good reason to doubt the supposed advantages of the planted-out over the pot-grown plants, all things being equal; but as soil and culture apply equally to both systems, growers may choose the method which suits them best.

In planting them out, strong well-rooted succession plants should be inserted in prepared beds of sound mellow loam, of a light or medium texture, with a free sprinkling of finely broken-up dry cow, sheep, deer, or pigeon dung—all are suitable in a *dry* state—soot, and bone-meal, making the soil quite firm around the plants. If the soil is in a proper state, no water will be required till the plants begin to push out fresh roots, when the soil will be in a condition to receive a thorough soaking, and no more should be given until the plants show to an experienced eye that they require it. If deluged unnecessarily with water, the soil sours and the plants grow soft and flabby in the leaves, and are difficult to get to show fruit, which is usually deficient in both size and quality. In consequence, liquid manure and rich top-dressings have to be used with the greatest caution; but a judicious application of these when the fruit is swelling is advantageous.

In pot-culture, which, with modern appliances, is the best for producing first-class fruit all the year round, similar soil is used, potting moderately firm, but not ramming it so " hard as pavement," as practised by some with no good result. For fruiting the Black Jamaica, Queen, and other varieties of moderate growth, 10 to 11-inch pots are quite sufficient for growing the largest plants to produce the finest fruits. For Smooth Cayenne, Charlotte Rothschild, and the like, 12 to 13-inch pots are ample; and even for the strongest growers, such as Black Prince and Providence, 13-inch pots, with proper attention, will produce finer fruit within a given time than pots of several sizes larger. All pots require to be well drained,

and the drainage covered with moss or similar material, to
keep the soil out of it, as nothing is so detrimental to success
as stagnant moisture at the root of a pine plant. To commence with, it is necessary to secure strong healthy suckers,
and from these first-class exhibition fruit may be had in about
eighteen months from the time they are inserted in the sucker
pots.   After the fruit is cut, if the suckers are not large
enough, they should be allowed to remain on the stool till
they are at least 18 inches in length, as, with plenty of water
and a growing atmosphere, they will make more growth in one
month on the stool than they will in three months in small
pots, if they are taken off before they are tolerably firm and
beginning to push roots from the base.   If room is required,
the stools can be removed and set close together in the succession or sucker pit, which suits them admirably.   As soon as the
suckers are ready they should be carefully removed from plants
bearing fruit, by placing the hand near the base and drawing
them off with a sharp twist, without injury to the leaves.   Then
put them into seven-inch pots in the same compost as already
described, making it firm around the base of the sucker. Place
them in a close pit, with the pots plunged to the rim in tanner's
spent bark or fresh leaves, and give no water for a week, or more
in dull cold weather, and only shade them from bright sunshine.
In about ten days the roots will be pushing freely, when the
plants should receive a thorough watering, and the shading
be gradually dispensed with.   The bottom-heat should range
about 85°, with an air temperature of about 70° at night,
rising 10° to 15° during the day, with a free circulation of air
whenever the thermometer touches 80°.   Pines require all the
light and air that can be given them in this country, and
thrive best when grown in a dry, clear, warm atmosphere,
mellowed with all the fresh air possible, so that the necessary
temperature is maintained.

In about three months, if all has gone on well, the pots
will be well filled with roots, and should be liberally treated
with clear liquid manure to maintain a vigorous growth till it
is convenient to shift them, which should be done within the
next six weeks, placing them at once in their fruiting pots, of

the sizes already mentioned, potting them firmly, and again plunging them to the rim in a bark bed. With three to four months of the same treatment—a due amount of heat at top and bottom, careful watering, a dry atmosphere, and giving all the light and air possible—the plants will have made fine sturdy growth and filled their pots with roots. They are now ready to be placed in the pine-stove or fruiting-house, and should be top-dressed with a rich compost of turfy loam, bone-meal, soot, and dry animal droppings rubbed down and sifted fine, all well mixed together and laid about an inch thick on the surface of the soil. Due attention must now be paid to watering, and as the pots are full of active roots, liquid manure should be given, which may be made from sheep, deer, or pigeon dung, soot, and guano, placed in a tub or old barrel filled up with boiling water, well stirred, and then allowed to cool and settle before being used. An occacasional sprinkling of a good fertiliser on the surface of the soil is also beneficial. The water should always be poured on to the surface of the soil, and not into the axils of the leaves, where it is not required. In about a year after the suckers have been put in, the plants will have attained a fine vigorous stocky size for fruiting, and by withholding water from them for three or four weeks—or about double that period in winter—and then giving them a thorough soaking with tepid water and slightly increasing the bottom-heat, they will immediately send up their fruit. The plants must now be regularly watered, but no stimulant given till each plant has done flowering, after which liquid manure or fertilisers may be applied at every watering with good results while the fruit is swelling. As the fruit nears maturity and begins to show colour, watering should be discontinued, as a plant in robust health will finish its fruit to perfection and of a brighter colour without farther stimulating. From ten days to a fortnight will ripen and colour a Pine-Apple in bright weather, but it may take a month in dull wintry weather. If a fruit is ripening too soon for the show, it may be cut a few days before it would be quite ripe and hung up in a dry fruit-room, where it will ripen and keep sound for two or three weeks without much loss in quality or freshness. The starting of the plants should, however, be so timed as to bring the fruit to full maturity and perfection as near to the date of the show as possible, as a fresh fruit is superior to a stale one. In proper pine-stoves, and under ordinary circumstances, Pines take about four months to reach maturity from the time the fruit appears till it is ripe; in summer a little less, but in winter two or three weeks more.

From the time the sucker is inserted till the fruit is ripe, the plants must be kept moving steadily, with the roots healthy and more or less active. If the roots become unhealthy from any cause, it seriously interferes with the fertility of the plant and the size and quality of the fruit, and special care must therefore be taken that they are in a good healthy medium, neither too dry nor too wet. When growing freely in spring and summer, a thorough watering once a week keeps them in perfect health. Towards the end of autumn they will not require it so often, and Queen Pines thrive best when they receive no water from October till they are started in December or January for ripe fruit in May and June. Black Jamaica, Smooth Cayenne, and others of the same nature require to be kept moister at the roots in winter, and every individual plant must then be closely watched, and water given to it only when it is necessary. By keeping the roots always healthy, the air dry and sweet, and giving all the light, heat, and air possible without causing cold draughts, Pine-Apples can be grown to the highest state of perfection and at the least cost. They are seldom grown specially for exhibition, but a few of the finest young fruits will well repay a little extra attention while they are swelling, when competition is in view.

It is usual to divide Pines into two or more classes in the prize lists of flower-shows,—the Queen and Smooth Cayenne each having a separate class, a distinction which their merits deserve, in the same manner as the Muscat of Alexandria and Black Hamburgh among grapes. A third class is generally provided for "any other variety," in which the large coarse varieties, such as Black Prince and Providence, frequently appear, to the exclusion of the smaller, but far richer, Black

Jamaica, the finest flavoured of all Pines, Charlotte Rothschild, St. Vincent, Ripley, and others of excellent quality.

The points of merit in a Pine-Apple are : (1.) Size, handsome shape, and a moderate-sized crown; (2.) rich flavour, ripe and sound, with flat, well-swelled pips; and (3.) colour, which should be the same from top to bottom, clear and fresh.

THE PLUM.

This is one of the most useful and productive of our hardy fruits, and is more cultivated in this country than any other fruit-tree, the apple alone excepted. It is a native of Asia Minor, but is naturalised in most of the temperate parts of Europe. In Britain it has been cultivated for its fruit for several centuries, and it has followed the spread of civilisation to all the temperate regions of the earth.

As an exhibition fruit, it fills a prominent place in a collection of hardy kinds, and a first-rate dish of Plums is a strong point in any collection of fruit. In competition, several classes are usually allotted to Plums, each having a well-defined character, such as the following :—

1. *Dessert Plums,* GAGES, some of the best of which for exhibition are :—Brahy's Green Gage, Decaisne, Green Gage, Oullin's Golden Gage, Reine Claude de Bavay, and Transparent Gage.
2. *Dessert Plums,* NOT GAGES ; among the best are :—Coe's Golden Drop, Denniston's Superb, Jefferson, Kirke's, Lawson's Golden, and Washington.
3. *Culinary Plums;* among the finest are : — Diamond, Goliath, Pond's Seedling, Prince Englebert, Victoria, and White Magnum Bonum.
4. *Damson Plums,* including all Plums under an inch in diameter, such as King and Prune Damsons, Burnet, Mirabelle, Mussel, and Quetche Plums.

The Plum is easily cultivated, and thrives well in almost any good soil well drained and of a warm nature. It succeeds in the open as a bush or a standard, but for exhibition purposes a wall is necessary to bring the finer varieties to perfection. Fan-training is the mode generally adopted, but where room for large trees is scarce, they may be grown as cordons with considerable success. Old trees should have their

N

branches freely thinned out, and young wood laid in at six or eight inches apart, as the best fruit is obtained from well-ripened young wood. The fruit should be well thinned as soon as it can be done with safety; neglect of which deteriorates the crop and spoils the prospects of another season. Liberal supplies of liquid manure and a sprinkling of some of the best fertilisers should be given in dry weather when the fruit is swelling, so that it may obtain a large size; and afterwards mulch the ground to preserve the roots, which run near the surface, and are liable to be injured in dry hot weather. All young wood not required to form the tree should be pinched at the third leaf, and the shoots wanted to fill the space carefully trained to the wall. When the fruit is ripening it should be fully exposed to the sun, so that its best qualities may be brought to perfection.

The points of merit in dessert plums are : (1.) Size; (2.) flavour; and (3.) colour. The same points of merit apply to the culinary varieties, although flavour is not of so much importance as size, thick flesh, and a small stone.

THE RASPBERRY.

The Raspberry belongs to the genus *Rubus,* and is a native of the temperate regions of Europe and Asia. It is very distinct in its habit, fruit, and the duration of its wood. The best of the red varieties are Antwerp, Baumforth's Seedling, Carter's Prolific, Fastolf, Northumberland Fillbasket, and Superlative; and of the yellow, Antwerp and Magnum Bonum. The Raspberry grows freely in the same soil as the Gooseberry, and enjoys a slight shade from the sun during the hottest part of the day. A position three feet from the north side of a not too high wall suits it well. The ground should be deeply trenched and well manured, and the canes planted three feet apart, placing three together, and afterwards cutting them down to one foot from the ground. By encouraging the growth of strong well-ripened canes, the plants should be established and able to bear good exhibition fruit the third year after planting. The old canes should be cut out in autumn, and the young ones tied in bundles of four in the form of an arch, or they may be tied singly to a trellis. The chief points in their culture for exhibition are to keep the canes thin, the surface roots intact, and to feed them liberally. The points of merit are: (1.) Size; (2.) flavour; and (3.) colour.

THE STRAWBERRY.

The Strawberry is a fruit that is widely cultivated, highly valued, and unique, in respect that its fine aroma and delicious flavour is appreciated by every one. It belongs to the genus *Fragaria*, some species of which are indigenous to Britain and Europe, while others are of American origin, which, by crossing and recrossing, have given rise to most of the varieties now cultivated. Some of the best kinds for exhibition are Elton Pine, Frogmore Late Pine, James Veitch, Lucas, President, and Sir Joseph Paxton. Certain varieties thrive and fruit well in almost any soil, while others refuse even to grow on the same ground; but as a rule they all succeed tolerably well on a loamy soil of medium character which is retentive of moisture but well drained. The ground should be trenched two feet deep, with plenty of well-decomposed manure incorporated among it. The best fruit is obtained from two year old plants, but excellent fruit may be gathered the first year after planting when that is done in August, especially if the young plants have been well rooted in small pots, and planted out of them with the balls intact. Three feet is a fair distance to give them between the rows, and 15 inches between the plants.

Watering the plants in dry weather and keeping the ground clean is all that is needed till early summer, when the runners should be removed as they grow, and the roots supplied with liquid manure after the fruit is set. In order to obtain fruit of the highest colour and flavour, the bunches should be thinned and kept well up among the foliage by means of a small hoop supported on stakes from six to eight inches in height. By this method of giving them all the light possible, the most difficult varieties to colour are brought to the highest perfection. In preparing for the second year's crop, which should be the most abundant, the same treatment should be given along with a surface dressing of rich manure, and some good artificial fertilisers to assist in perfecting a heavy crop. Strawberries are largely grown and exhibited by amateurs, who tend them with the greatest care, and watch the development of the fruit with as keen interest as they do the opening flower of the seedling pansy, and no care or trouble is spared to secure a favourable verdict to the object of their choice.

The forcing of strawberries for spring and early summer shows is a process requiring skill and care to ensure success. Some of the best varieties for this purpose are Auguste Nicaise, James Veitch, Keen's Seedling, La Grosse Sucrée, President, and Vicomtesse Héricart de Thury. Strong runners should be layered as early in the season as possible, in small pots, among sandy loam and leaf-mould. As soon as they are well rooted, repot into six-inch pots, using a compost of old fibry loam and a fourth of well-rotted hot-bed manure, to which a sprinkling of bone-meal, sharp sand, and wood-ashes may be added. Set the plants in rows in an open space, protect the pots from strong sun, and to promote strong growth give occasional waterings with weak liquid manure. Give them protection in winter from hard frost, and keep the soil moist to preserve the foliage as long as possible. Introduce them to a temperature of 50° at night about three months before the fruit is wanted; keep them close to the glass, and raise the heat to 60° as the season advances. Thin out, and stake the flower trusses, give plenty of air in favourable weather, and liquid manure to assist in swelling the fruit. Strong crowns, slowly forced, with plenty of air, light, and moisture, with judicious feeding, should produce a crop of fine fruit, which early in the season is highly valued. The points of merit are: (1.) Size and form; (2.) flavour; and (3.) colour.

DIVISION IV.

VEGETABLES.

THIS is by no means the least important division in an exhibition of horticultural produce, and although it is not so showy and generally attractive as the flowers and fruits, it affords much interest and instruction to the public, and is a field for the keenest competition among horticulturists. The knowledge of how to grow vegetables to perfection is of the utmost importance to all who cultivate a bit of ground, from the smallest allotment of the cottager to the extensive culinary department of a county magnate's garden establishment, where the very best of everything is a *sine quâ non;* and although the following directions are specially designed for growing specimens for exhibition, they are equally applicable to the cultivation of ordinary vegetables. In mentioning the best varieties of the different classes of vegetables, only well-known types, as a rule, have been selected, because it is scarcely possible to name varieties which would do equally well under all circumstances in every part of the country, but they may afford a guide to the inexperienced beginner and the amateur. All experienced competitors have their own select varieties, which they cultivate with indomitable perseverance and skill, to secure the highest awards in competition.

THE GLOBE ARTICHOKE.

The varieties of this vegetable in cultivation are the Green and Purple with different strains from each of more or less merit. To procure first-rate fleshy heads of the Globe Artichoke, a strong rich soil and an open situation are required. Plants should be divided and planted three feet apart in fresh soil. All the small shoots should be cut away, and two or three strong crowns left to each plant. Protection should be afforded to the plants in winter by placing dry leaves and litter around them, which should be removed in spring, and a good dressing of rotted manure forked into the soil without disturbing the roots. As soon as the heads are formed they should be thinned, leaving one on each stalk. The plants should then be watered with liquid manure, to increase the size of the head. They should be exhibited when fully developed, but close and firm. The points of merit are : (1.) Size of head; (2.) substance of the bracts; (3.) colour (green preferred) and fitness for use.

THE JERUSALEM ARTICHOKE.

This is an excellent winter vegetable, which can be appropriately exhibited in either autumn or spring. It should not be cultivated on strong rich ground, because the top growth attains great dimensions, while the roots are generally small and badly formed. A rather sandy soil should be chosen in a position sheltered from wind, but well exposed to the sun. The best-formed medium-sized roots should be planted during March in rows three feet apart and 18 inches between the sets. Under these conditions, and with frequent drenchings of liquid manure when the tubers are swelling, large, clean, well-shaped roots should be formed. When the stems have reached four to five feet in height, they should be stopped, and the side-shoots kept close in by pinching. The points of merit are : (1.) Size; (2.) form ; and (3.) an even sample of tubers.

THE ASPARAGUS.

The Asparagus cannot be grown successfully in every kind of soil. Where the natural soil is unsuitable a bed should be made up of rich friable loam and well-decomposed manure. Two of the best for exhibition are the Giant and Connover's Colossal. To secure large heads, the plants should be thin on the ground. A heavy dressing of salt should be given to the beds in spring before the heads appear, and sprinklings of guano and nitrate of potash alternately in moist weather during growth are very beneficial. Cutting should cease by the middle of June at latest, after which the strongest shoots should be carefully preserved, and staked if necessary, to prevent them being swayed about and broken by the wind, the weak ones being cut away as they appear. As soon as the shoots are ripe in autumn they should be cut over, and the ground covered with six inches of stable litter and leaves. The points of merit in Asparagus are: (1.) Length and thickness of the stalk; (2.) the head close; and (3.) all of an equal size and fit for use.

THE BROAD BEAN.

Three of the best varieties of Broad Beans for exhibition are Aquadulce, Leviathan, and Seville. They require a rich stiff soil, well worked and pulverised, to grow in. Seed should be sown at three or four different times between the beginning of February and the end of May, in drills five inches deep and two feet apart. Earth should be drawn up to the tops of the earliest ones when they appear above ground, to protect them from frosty winds. When the plants have reached the height of four feet they should be topped, and the pods thinned to five or six of the best formed, equally divided up the stem. A mulching should be given to them after being watered with weak liquid manure, an operation which should be attended to often in dry weather. By careful thinning and feeding, pods of great length, containing extra large beans, should be produced. The points of merit are: (1.) Length and breadth of pods, which should be all equal and dark green in colour; (2.) number and size of beans, which should be in the best condition for use.

THE FRENCH BEAN.

A high-class vegetable, and when in its most perfect state about equal to the pea as an exhibition dish. The best varieties for that purpose are Canadian Wonder and Ne Plus Ultra. If they are required to be grown under glass, the latter is the best for exhibition. French Beans thrive in a rich warm soil, and to have them in their best condition, they should be sown in prepared trenches similar to peas. The first sowing should be made early in May, and be followed by others at intervals of a fortnight till the middle of June. The beans should be planted singly, at about $4\frac{1}{2}$ inches apart, and half of them pulled up when all are safe above ground, giving each plant left nine inches in the row, with $2\frac{1}{2}$ feet between the rows. The plants should receive a copious supply of water in dry weather, giving weak liquid manure when the pods are swelling. If the pods are too forward for the show, they must be removed before they attain full size, so that those coming in at the proper time may receive the full benefit of the vigour of the plant.

The points of merit are: (1.) Size; (2.) quality, young and tender, and showing no fibre when broken across; and (3.) appearance, smooth and equal in length and shape, and of a bright green colour.

THE RUNNER BEAN.

Two fine exhibition varieties are Jubilee and Scarlet Champion. The cultivation of the runner is similar to that recommended for the general crop of French Beans, except that long stakes or a trellis should be provided for their support. The pods should be thinned out, and never be allowed to get old on the plant. The points of merit are the same as those in the French Beans.

THE BEET.

There are many varieties of Beet, each raiser claiming superiority for his special production. Carter's Perfection, Pragnell's Exhibition, and Frisby's Excelsior are among the best for exhibition. Beet requires a light soil, free from vegetable matter, to grow in, with some rich manure about a foot from the surface. Seed should be sown in April and May, in drills 18 inches, and the plants thinned to nine inches apart. The points of merit are: (1.) Size and smoothness of the root; (2.) flesh crisp, free from fibre, and when cut the sap should "bleed" or ooze; and (3.) colour dark with the circles wide apart, and all of the same colour.

THE BORECOLE OR KALE.

The Kale is a winter vegetable of much importance, especially in the northern parts of the kingdom. There are many varieties of the dwarf curled, each grower selecting and naming his own strain, which is the best way to ensure its being kept true. The cultivation of Borecole is extremely simple; it only requires a rich soil and plenty of room. Seed should be sown in a frame in the beginning of March for an early supply, but for autumn and winter a sowing should be made on a sheltered border in the middle of March, and the seedlings thinned early to prevent the plants being drawn. As soon as the plants are four inches high, plant out two feet apart each way. Water in extra dry weather, giving weak liquid manure occasionally to increase the size of the head. The plants should be cut clean across below the leaves, and the stems stuck in pots amongst wet sand for exhibition. The points of merit are: (1.) Size of head and number of leaves, the older ones slightly arching; and (2.) the leaves dark or light green, succulent, and densely curled.

THE BROCCOLI.

The many varieties, both early and late, of this fine vegetable furnish a long succession of heads for autumn, winter, and spring exhibitions. To provide a supply for each of these periods, seeds of Walcheren and Veitch's Self-Protecting should be sown thinly in April to produce heads for the autumn; *Backhouse's Winter* and *Snow's Superb* for winter, and *Niddrie Protecting* and *Veitch's Model* for spring and early summer; sowing these at intervals of two weeks from the middle of April till the end of May. After the plants are large enough to handle, they should be pricked out four inches apart on a bed of fine soil and leaf-mould until they are large enough to be finally planted out.

Broccoli thrives best in a rich loam, which should be deeply dug and well trodden to firm it. Previous to planting, a dressing of salt and soot should be pointed into the surface for the destruction of maggot and other pests.

The plants should be lifted from the bed with balls of earth, planted in rows three feet apart and two feet between the plants, and watered if the soil is dry.   The space between the rows should be occupied with spinach, lettuce, or turnips where ground is scarce.   The advantage of wide planting is observed after a cold winter has passed.   The autumn and spring Broccoli should be assisted to make large heads after they are formed, by frequent applications of liquid manure. When the heads appear, they should be protected from light and frost.   When cut with a long stem, they can be preserved for two weeks in a dark cellar, and the colour improved by the keeping.

The points of merit are: (1.) Size and symmetry of the head ; and (2.) closeness and pureness of the "curd," which should be free from "froth."

THE BRUSSELS SPROUT.

There are many so-called varieties of this fine vegetable, but it is rare to find them true to description, from the want of sufficient care being taken in the selection of the plants from which the seed was obtained.   A grower for competition should select a few of his best plants every third year, lift and plant them at the foot of a wall as far away as possible from other flowering plants of the brassica tribe.   When the seed is ripe, it should be rubbed out, preserved dry, and sown on a gentle hot-bed during the first week in March.   The seedlings should be pricked out and hardened off by the end of April, when they should be planted out two feet apart on well-prepared ground, which is inclined to be heavy rather than light.   Liquid manure should be given after the sprouts are formed, and the plants never allowed to want for water during dry weather.   The plants should be lifted when wanted, the lower leaves trimmed neatly off, and then potted with a few of the roots attached.   By this method the plants are kept fresh, and their parts better exposed than when they are tied together in bundles.   When sprouts are exhibited as a dish, the best only, and all of equal size, should be selected.

The points of merit are : (1.) Size and number of sprouts on the stem ; (2.) equality in size ; and (3.) closeness and firmness of the sprouts.

THE CABBAGE.

Although one of the commonest of all vegetables, the Cabbage is by no means an unimportant item in an exhibition collection, when it is set up in its very best condition. The best variety for early shows is a good strain of Early York, and the best for autumn shows is Winningstadt, with any good strain of Red Cabbage for pickling.

The Cabbage thrives best in a strong rich loam, and in inferior soils it must receive extra liberal treatment to make up the deficiency. Seed of Early York should be sown on a warm border about the middle of August, and again a fortnight later. Plants from these sowings will be ready to set out in October and the following February. Seed of both Early York and Winningstadt should be sown in the same manner in February, and every fortnight till May, to keep up a succession of fresh succulent heads during the exhibition season. The same treatment as applied to the Cauliflower suits the Cabbage admirably. The points of merit are : (1.) Quality, the leaves should be thick and succulent, and closely turned in ; (2.) form, which should be more or less conical, not flat-topped nor oblong shaped ; (3.) Size, firmness, and weight.

THE CARROT.

When clean and well grown, the Carrot is an important dish in a first-rate collection of vegetables. It grows to great perfection in certain soils without any trouble, while in others it completely baffles the best skill to make it thrive. The most certain method to command success is to provide the proper soil for it. The best varieties for exhibition are Early Horn, Nantes, and a good strain of Intermediate. Clean sharp sand from the sea-shore is the best medium in which to grow carrots to perfection, but if it cannot be easily obtained, river or pit sand will do. It should be placed in a wide trench or bed

15 to 18 inches in depth, and enriched with dressings of salt and liquid manure poured over it during the winter. Fork it over loosely before sowing the seed in shallow drills, 15 inches apart, early in April. Thin the plants in due time to six or eight inches apart, according to the variety, and keep them always clear of weeds. Apply weak liquid manure in dry weather, and sprinkle guano or any good artificial manure among them on showery days. The points of merit are: (1.) Size, which should be large of its kind; (2.) form, which should be smooth and symmetrical; and (3.) quality, the flesh should be crisp and tender, bright red in colour, and the core small.

THE CAULIFLOWER.

While Cauliflower is in season it is indispensable in a collection, and when in its best condition takes a leading place among vegetables. The best varieties for exhibition in succession are Snowball, Eclipse, and Autumn Giant.

If required early, a sowing should be made in boxes in a gentle heat about the end of February, and care must be taken that the young plants do not become drawn, or want for water while they are under glass. When large enough to handle, they should be pricked off at three inches apart in boxes of rich soil. Gradually harden off, and plant out as soon as all danger from frost is past, always being careful that the plants receive no check from any cause, which is apt to make them " button." Successional sowings must be made in the open air about every fortnight from March to May, and the plants treated as carefully as those raised under glass. A strong loam, heavily dressed with well-made manure, grows the finest Cauliflower, and where the soil is light and poor, special care should be taken to improve its quality, by adding good loam and an extra quantity of manure. When ready for planting out, the young plants should be carefully lifted with a good ball around the roots, and planted with a trowel, so as to injure them as little as possible. Settle the soil around them and give a good watering. Repeat the watering every few days if the weather is dry, and when the heads begin to form, liquid manure should be freely applied to develop them to their finest proportions. As the heads expand, be careful to shade them from the sun by bending the leaves closely over them, to exclude the light and preserve their pure white colour. Should they come too early for the show, they may be preserved for a short time in a tolerably fresh state by being pulled up by the roots, and hung, heads down, in a dark cool place. In judging, the points of merit are : (1.) Form, which should be globular and regular in outline ; (2.) size ; and (3.) colour, which should be pure white.

THE CELERY.

This vegetable in its best form is one of the strongest articles in a collection, and as a single dish it takes first rank, especially when produced, solid and crisp, in early autumn, 18 inches in length of blanched stalks, and 15 inches round them. Wright's Giant White and Manchester Red are excellent varieties for exhibition. Seed should be sown in boxes in the middle of February, and placed in a mild hotbed, or in a pit with a night temperature of 55°. As soon as they are fit to handle, they should be pricked out, about four inches apart, on a gentle hotbed or in boxes, using rich friable loam, and being careful never to allow the plants to become dry at the root, which is exceedingly prejudicial. Towards the end of April the plants should be nice stocky stuff, and should be gradually exposed to the air by pulling the sashes off the frame more or less daily according to the weather, until they are quite hardened off, and **ready** for planting out in the trenches.

For growing exhibition celery, a sheltered place lying in the full sun should be chosen. The ground should be well drained, naturally or artificially, and the trenches prepared some time before they are required. For a single row of plants, the trench should be 15 inches wide and 18 inches deep, removing any bad soil that may be turned up, and filling the trench with one foot in depth of the following compost: one-half sound turfy loam, and the other half of equal parts of well-rotted manure, leaf-mould, night-soil, wood-ashes, and sand. Carefully lift the plants with a hand-fork, to preserve the ball as whole as possible, and plant them in the trench at one foot apart. Press the earth lightly around them, and then give a thorough soaking of water to settle the soil around the roots. After the plants are dry, place a piece of thick brown paper around the stalks of each (leaving the tops free), which will support them and prevent the soil getting in the heart of the plant, and assist in the blanching of the stalks. A little earth should be drawn up around the paper to keep it in position. If the weather is dry, copious waterings are necessary to ensure a free firm growth. In a few weeks they will be growing freely, and will require earthing up. The paper "collar" should be slipped up, and the stalks being all in proper position, a little dry sawdust or sand should be placed closely around them, and banked up with the earth dug out of the trench. When the plants are growing freely, weak liquid manure and top-dressings of artificial manures are beneficial; but discretion is required not to overdo the use of stimulants. Manures in excess produce coarse rank stalks, worthless, as a rule, for exhibition purposes. Good rich soil, with a moderate amount of manure, and a judiciously regulated supply of liquid and artificial manures during growth, produce the best results. In dry warm weather the plants are refreshed by a sprinkling from a watering-pot in the evening, scattering a little common salt among the plants previous to the sprinkling. Earthing should be carefully attended to till the stems are blanched to the desired length, which, however, should never be obtained at the sacrifice of girth, the first object being thick, solid, crisp stalks, and then length of blanched stem.

The points of merit are: (1.) Size (length and girth) of blanched portion; (2.) quality (solidity and crispness); (3.) flavour, a rich "nutty" flavour being preferred. No "bolted" celery should receive an award.

THE CUCUMBER.

Among the best varieties of the Cucumber for exhibition are Duke of Edinburgh, Telegraph, Tender and True, and Verdant Green. They are easily grown in a glass frame set on a hotbed, where fine specimens may be obtained with the aid of glass tubes to keep them straight and shapely. The best method, however, of obtaining clean, straight, and handsome specimens is by growing them in a properly constructed cucumber-house, where the plants are trained on wires near the roof, and the Cucumbers hang down from them, clear of all contact, and fully exposed to plenty of light and air. Seed should be sown in light rich soil, in a high temperature from ten to twelve weeks before the produce is wanted; and as soon as the plants are in the rough leaf, the seedlings should be potted singly and grown till they are about a foot high, when they should be planted out on small hills of light fibry loam and leaf-mould, and the house kept at a temperature of 70° at night, with abundance of moisture in the atmosphere, and shading from sunshine when necessary. The plants should grow quickly and show fruit within three weeks, which should be thinned to a few of the most promising which are likely to be at their best on the date of the show. The growth must be kept well thinned out, and the bearing shoots stopped at the leaf beyond the fruit. Those coming too early should be removed in time, to conserve the energies of the plant for the production of the exhibition specimens, which are much benefited by frequent doses of liquid manure when they are swelling. For exhibition, fruits of the same size and appearance should be chosen; and in handling and packing, every care should be taken to preserve the bloom intact. The points of merit are: (1.) Equal size and shape; (2.) fresh and dark-green in colour; and (3.) the bloom perfect.

THE ENDIVE.

There are many varieties of this useful vegetable, all raised from the broad and curled leaved. The latter makes the finest-looking head for exhibition, and should be cultivated for that purpose. Seeds of both the green and white curled should be own on a rich border in May and in June for succession. As soon as the plants are large enough, they should be transplanted nto rows 18 inches apart, and one foot between the plants, upon a deep free soil that has been previously well manured. To promote rapid growth for the production of tender succulent leaves, frequent waterings should be given as soon as the plants get established, using drainings of the manure heap alternately with clear soft water. When the plants are about full grown, the blanching should be proceeded with, while watering with liquid manure should cease. The simplest mode of blanching is by placing pots or seakale covers over the plants, which should be done about three weeks before the heads are wanted, when the plants should be carefully lifted and put into 7-inch pots for exhibition. The points of merit are : (1.) Size of head ; (2.) closeness of heart, with leaves well curled ; and (3.) well blanched and tender.

THE HORSE-RADISH.

This useful perennial presents no choice of varieties, but possesses within itself, when well grown, all the qualities that can be desired for the purposes for which it is used. To obtain large clean roots for exhibition, a trench 18 inches wide and deep should be dug in March, into which six inches of well-rotted manure should be dug. The surface soil should then be replaced, and holes made one foot deep and the same apart, into which strong clean pieces 12 inches long with the crown left on should be dropped. The soil should then be firmed and watered if dry. The roots should be lifted when they are wanted, the leaves and small roots trimmed off, washed, and tied in bundles for exhibition. The points of merit are : (1.) Length and thickness ; and (2.) straight, clean roots.

THE LEEK.

When grown in its best form, the Leek forms a conspicuous feature in a collection of vegetables. The best varieties for exhibition are The Lyon and Renton's Monarch. Leeks for exhibition should be sown in heat about the middle of February. The best method is to fill as many $3\frac{1}{2}$-inch pots as may be required with light rich soil, in which about half-a-dozen seeds should be placed. When about three inches high, the best plant should be selected, and all the others pulled out. As soon as the roots fill the soil, the plants should be transferred to six-inch pots, and grown on in a night temperature of 55°. Early in May they should be gradually hardened off, preparatory to being planted out. Trenches should be prepared for them in the same manner and with the same materials as for Celery, and after planting out they require exactly the same treatment—watering, top-dressing, and earthing-up—to bring them to the highest perfection. The points of merit are: (1.) Size or length and girth of blanched portion; (2.) shape, which should be cylindrical, without any bulge at the root; and (3.) firmness and weight.

THE LETTUCE.

This is an indispensable vegetable in all gardens, and is the most generally cultivated of all the salad plants. There are two well-known and distinct types in cultivation, the Cabbage or round Lettuce, and the Cos or upright form, the latter being the favourite for exhibition. A host of varieties are grown, but a good strain of the White Cos is the best to grow for competition specimens. Seed sown on an open border in March, and planted out as soon as large enough, will furnish fine samples by midsummer, or they may be brought on much quicker if grown in a close frame or under hand-lights. For raising early and tender Lettuce, the French *cloche* is a first-rate contrivance, but it is more suitable for the Cabbage varieties—the favourite in France—than for the Cos; still, with its aid, the quality of the latter is much improved. A rich friable soil, well dug and manured the previous autumn, is a first-rate medium for growing the finest of Lettuce. When they are about half-grown, they should be well fed with liquid manure, and receive abundance of water in dry weather. Attention must be paid to tying the heads with a piece of soft raffia to blanch the leaves, if they do not close in naturally. A Lettuce for exhibition should be neatly dressed by removing the root and all stale or damaged leaves, leaving nothing but what is fit to make a salad. The points of merit are: (1.) Size and solidity of the head; (2.) quality, fresh, crisp, and tender; and (3.) leaves thick and juicy, and of good flavour.

THE MUSHROOM.

The Mushroom is much appreciated as an esculent, and consequently attains a high position among others in competition. Its cultivation is chiefly confined to places of importance and to growers for market, and many methods have been devised for its production on a cheap scale, on ridges and in frames without artificial heat. These modes are sometimes very successful; but a house specially constructed for the purpose, and where the necessary conditions for growth can be maintained, independent of external circumstances, is the best and most reliable. Mushrooms cannot, like other vegetables, be much stimulated with liquid manure during their time of growth, consequently the bed on which they are to grow should be made of the most substantial material. Horse-droppings, turned daily to sweeten for about a fortnight, are generally used for this purpose, but a mixture of cow and sheep droppings tend to modify the heat and render the material far more lasting, giving larger produce and of a better quality. These materials, after being mixed and two or three times turned, should be put in a bed about 18 inches deep, and made very firm. In a few days the heat will rise, and after it begins to decline the spawn should be introduced. A layer two inches thick of loam and cow-dung made into mortar, should then be spread over the bed and made smooth. A moist atmosphere should be constantly maintained and the surface of the bed covered with straw to keep it moist. Little or no fire-heat will be required in summer, but in autumn and winter a steady temperature of 56° should be maintained. The points of merit are: (1.) Size, which should be equal; (2.) condition, fresh and perfect; (3.) colour, white with the gills a fresh pink.

THE ONION.

The improvement in the cultivation of the Onion has perhaps been more marked in recent years than that of any similar vegetable. This is partly due to the introduction of improved varieties, and particularly the large varieties, such as Trebons, Giant Rocca, and the like. The size which these and others now assume was quite unknown in onion-culture a few years ago; and still the size increases yearly, our leading exhibitors vieing with each other in "beating the record." Three of the finest for exhibition are Ailsa Craig, Rousham Park Hero, and Trebons. Onions thrive best in a rich, light, warm soil, made firm on the surface by treading just before planting. It should be heavily manured and double dug the previous autumn, leaving it in ridges exposed to the frost all winter. When required for planting, level the ridges and fork into the surface soil a dressing of salt and soot. A little bone-meal is also very beneficial. The seed may be sown in the open air on a warm border about the middle of August, or in boxes, in a mild heat, early in February. In either case they must be pricked three inches apart into boxes, filled with light rich soil, and placed in a temperature of about 55° at the lowest, where they will grow freely and make fine plants to set out in due time. Care must be taken to prevent any drawing of the plants while they are in heat, and they must be gradually hardened off to be ready to plant out early in May. The ground being properly prepared, the planting should be done on a mild afternoon, lifting each plant carefully with a ball, and inserting it unbroken into the firm surface of the ground with a trowel. Place a small stake to each plant, to save the leaves from being tossed about by the wind till the plants are established, and give a watering to settle the soil around the ball. The rows should be about 15 inches between, and the plants about nine inches apart in the rows. They should be kept clear of weeds, and receive when growing freely a copious supply of liquid manure (made from guano, soot, or pigeon-dung) in dry weather, and a sprinkling of artificial manure in showery weather is very beneficial in swelling the bulbs to their largest dimensions. As they reach maturity the heads of most of them require to be bent over and slightly twisted, to stop top growth and assist the ripening of the bulbs. For early autumn shows it is generally necessary to lift them and twist their necks, and place them in a dry hot place to ripen for competition. Their points of merit are: (1.) Size, which must be firm and weighty; (2.) form, which should be globular and smooth, with a fine neck; and (3.) ripeness.

THE PARSLEY.

This vegetable, when shown as part of a collection or in a class by itself, should always be exhibited in a pot. It should not be simply lifted and potted when required, but it should be grown at least for a time in the pot. Two suitable varieties are Myatt's and Triple Curled. Seed should be sown in a box in March amongst light loam and leaf-mould, and placed in a frame. When the plants are large enough to handle, they should be pricked four inches apart into another box. After the plants are strong and bushy, they should be potted into six-inch pots among similar soil, and set in a frame with a northern aspect. They should be shaded from strong sun, and never be allowed to get dry. A moist atmosphere around the pots is also essential to promote healthy growth. Weak guano-and-soot water, given once a week after the plants are established in the pots, encourages the production of large healthy leaves. The points of merit are : (1.) Size and number of leaves ; (2.) quality, fresh and tender ; and (3.) plant well furnished with firm, densely-curled leaves.

THE PARSNIP.

The varieties of parsnips are few, two of the best being Hollow-Crowned and The Student. To secure long, clean, tapering roots, parsnips should be cultivated on a deep sandy loam free from all rank vegetable matter. The ground should be prepared during winter by trenching two feet deep, and digging into the bottom of the trench a quantity of rich decomposed manure. The ground should be well broken up in March, and the seed sown thinly in rows, two feet apart. The plants should be thinned to one foot apart, and when the roots are about half grown and well formed, liquid manure should be given, which will do much to feed and increase the size of the root. A deep trench should be dug between the rows when lifting, and all the root that is clean and straight preserved. The roots should then be washed clean, and all side fibres nipped clear away. The points of merit are : (1.) Length and thickness ; (2.) root tapering, straight, and clean ; and (3.) flesh firm, tender, and succulent.

THE PEA.

This is one of the most important of all vegetables. In competition, a dish of peas in a perfect condition takes a high place. Three of the best varieties for exhibition are Duke of Albany, Stratagem, and Telephone. To grow it to perfection, the Pea requires a strong rich loam. Where the soil is un-favourable, trenches should be filled with a mixture of turfy loam and well-rotted manure six inches deep, and the Peas sown thinly on it, covering them with an inch deep of ordinary soil. For exhibition in August and September, two sowings should be made in May. As soon as they are fairly above the ground they must be properly staked, and if the weather is dry, they should be mulched with long litter and regularly watered. When they have grown to nearly their full height, the tops may be pinched off. If the crop is heavy, it should be thinned by removing the small and badly formed pods, and also all that are likely to be too forward. By this means, fresh, plump, well-filled pods are secured. Peas should be shown in their pods, and a certain number of pods should always be stipulated for. The points of merit are : (1.) Length and girth of pod, which should be straight and well filled, and all of an equal size ; and (2.) Peas large, plump, dark-green, tender, and sweet to the taste.

THE POTATO.

The varieties of this useful esculent have increased very rapidly within the last few years. They have also been very much improved in appearance, but whether or not the quality has improved in the same degree cannot be definitely stated. We are, however, indebted to those specialists who by careful cross-fertilisation have been enabled to produce varieties, proof, at least for a time, against the attacks of disease.

Among the best twelve for exhibition are Ashleaf Fluke, Beauty of Hebron, Excelsior, Freedom, International, King of Russets, Schoolmaster, Snowflake, Snowdrop, The Cobbler, Vicar of Laleham, and Village Blacksmith. White and round varieties being most popular, should predominate in all collections.

The potato is of easy culture, but it cannot be successfully grown for exhibition in every kind of soil. Certain varieties grow well in a moist situation, but the majority produce the best tubers in a dry soil. A light mellow loam, comparatively free from iron, and well mixed with leaf-soil, is most suitable for the production of clean well-shaped tubers.

To secure late potatoes for early autumn shows, it is necessary to start the sets in heat in the middle of February, placing some of the best-formed medium-sized tubers into shallow boxes amongst leaf-soil. As soon as they have grown an inch, they should be removed to a cold frame and placed near the glass to harden off. The soil having been well broken up and in a workable condition, they should be planted out in the middle of March. Lift them out of the boxes with as much soil attached to the roots as possible, and place them in shallow furrows from two to three feet apart, and one foot to one foot four inches between the sets, according to their habit of growth. All the shoots should be removed from the sets at planting except the strongest one. A little fine soil should be placed · round the sets in the drill, then cover them to a depth of four inches. Potatoes should not be planted deep, because the best tubers are invariably those nearest the surface. When the tops appear above ground, a little earth should be drawn over them for protection. The hoe and fork should be often applied between the drills whilst the potatoes are young, as the more they receive in this way, especially in adhesive soil, the better will be the crop. Previous to earthing up, a dressing should be given of some inorganic manure recommended for the potato, it being more beneficial to them than quantities of rank farm-yard manure. The soil should be drawn well up to the stems when they are finally earthed up, and all blooms picked off as they appear.

Care should be exercised when lifting the crop to prevent injury to the skin of the tubers. The best specimens should be picked out as they are lifted and set apart, and in the final selection for exhibition, those of equal size, form, and good appearance should be chosen ; and after being carefully washed, they should be rolled in soft paper and kept from the light.

When dishes are not provided for staging, a shallow box three inches deep is necessary for the purpose, into which the tubers should be laid upon green moss, arranging the colours tastefully, and keeping an inch of clear space between each tuber. The points of merit are: (1.) The tubers should be of a good size, equal and alike in form and colour; and (2.) the flesh when cut across should be clear and of one colour. Coloured varieties with a pink tinge in the flesh should not be disqualified on account of it.

THE RADISH.

There are numerous varieties of the Radish in cultivation, the best of which for exhibition are the French Breakfast and Long Scarlet for forcing and the earliest crops in the open air, and the turnip-rooted red and white varieties for summer and autumn crops. They thrive best in a rich, warm, free soil, in a sheltered spot in spring and early summer, and on a north sloping border in hot summer weather, when the direct rays of hot sunshine are detrimental to their quality. Seed should be sown from four to six weeks, according to the season, before the date when the roots are required. If the weather proves hot and dry, they are much benefited by a plentiful supply of water and liquid manure to promote a free and rapid growth, upon which their chief qualities depend. In selecting them for exhibition, choose clean well-grown roots, which break crisp and clean, and show no fibre in them. The points of merit are: (1.) Size, and all the sample equal; (2.) quality, fresh and clean from small fibry roots; and (3.) flesh crisp and juicy, and of a brisk, pleasant, slightly pungent flavour.

THE RHUBARB.

This accommodating vegetable may be said to grow anywhere, but can scarcely be placed in any soil too rich for its taste. Two first-rate varieties are Johnstone's St. Martin and Victoria. The former is comparatively sweet, while they both possess a fine red colour, and can be grown to a large size. The open-air culture being simple, a few hints on forcing of Rhubarb may be useful. The produce from plants lifted and

forced being inferior for exhibition to that from those established outside, a few plants should be divided in March, and
the strongest crowns planted three feet apart in rows five feet
asunder, on a deeply-trenched, well-manured, warm border.
When the plants begin to grow, they should receive frequent
waterings with drainings from the stable or byre, well diluted.
No stalks should be taken from the plants during the summer,
that the whole strength and substance may be concentrated
for their future support. Square boxes, three feet wide at the
bottom and two feet at the top, should be put over the plants,
around which should be built a good thickness of stable litter
and leaves well mixed. A little light admitted at the top
during fine days improves the colour and texture of the stalks.
The points of merit are: (1.) Length and thickness of the stalks;
(2.) quality, firm, tender, and briskly acid; and (3.) colour,
clear and bright.

THE SAVOY.

The Savoy is an excellent winter vegetable, which, like the
Brussels Sprout, is the better for a little touch of frost to make
it tender. Good varieties for exhibition are Early Ulm, Green
Curled, and Drumhead. When Savoys are wanted for autumn
seed should be sown under the same conditions as are recommended for Brussels Sprouts, while the second sowing should be
made along with the Cabbage on an early border. The plants
should be afterwards treated in the same manner as Cabbage.
The points of merit are: (1.) Size and firmness of the head;
(2.) solidity of the heart when cut; and (3.) yellowish colour
of heart.

THE SEAKALE.

This important winter and spring vegetable thrives best in
light deep loam, and the best heads are reared from two-year
old plants. Seakale should be treated in the same manner as
Rhubarb, except that a little salt should be sprinkled round
the plants previous to watering, and blanching covers put over
the crowns instead of boxes. The growth of the shoots can be
advanced or retarded by the strength of heating material used.
The points of merit are: (1.) Size; (2.) compactness; and
(3.) quality, crisp, succulent, and perfectly blanched.

THE TOMATO.

The Tomato has of late become exceedingly popular, and although it takes a place among vegetables, it is very often used as a fruit, and may some day be included in that class. It makes a very bright and attractive dish, and in a collection of vegetables it takes a high position. Many new varieties have been recently sent out, some of which are good, while others are inferior to some of the old sorts. Hathaway's Excelsior, Large Red, and Trophy are excellent old varieties, seeds of which should be saved from the largest and finest of the produce. And of newer varieties, Dwarf Champion, Hackwood Park, and Volunteer are among the best. They are very easily cultivated if a low-roofed house with plenty of space for ventilation and means for maintaining a night temperature of 50° to 55° is provided for them. To produce ripe Tomatoes in June, seed should be sown in a mild heat about the beginning of March, and the seedlings potted on till they fill five-inch pots, after which they should be planted two feet apart on a bed composed of light loam and leaf-mould. The plants should be led up with a single stem, and all laterals pinched out as they appear. As soon as they come into flower, a free circulation of dry air through the plants will ensure a good set, and as soon as the fruit are formed, all bad-shaped ones should be cut out, and one or two of the best left on a bunch. To produce large fine Tomatoes, the time to feed is after a number are set and swelling. If it is done with liquid manure or rich soil previous to this time, soft growth will result, and unsatisfactory exhibition produce will be obtained. We use at this period a heavy top-dressing of loam and well-decomposed manure in equal parts, to the barrowful of which is added a six-inch potful of a good fertiliser and the same of bone-meal, all mixed together, spread, and made very firm over the roots. As soon as the roots are well through the soil, the plants are supplied with liquid manure once a week. The effect of this treatment is fine Tomatoes and a long season of produce. The points of merit are : (1.) Size ; (2.) form ; and (3.) colour. The fruits should be of equal size, regular in outline, and of a bright clear colour.

THE TURNIP.

The Turnip is an indispensable vegetable in a mixed competitive collection. When a class is provided for itself, it is generally well filled, and the exhibits in many cases reach the standard of perfection. Two of the best for early shows are Snowball and White Model, while a good strain of Golden Ball is the best for autumn and winter competition. To grow turnips of fine form, free from coarse fibres and with small roots, seeds should be sown between the middle of May and the end of June on rather poor, light, loamy soil. But the roots must not be starved, else the flesh will be tough; consequently the earth should be drawn well over the bulbs when about half grown, and a furrow made from which to feed the roots with weak liquid manure without its coming in contact with the bulbs. Bulbs of a medium size are preferred to extra large ones. They should be lifted a day or two before the show and washed, the rough parts rubbed smooth, and any fibres removed; after which the yellow kinds should be exposed to the light, which improves the colour. The points of merit are: (1.) Form; (2.) colour; (3.) size. The white should be pure, and the yellow kinds bright and uniform all over. When cut, the flesh should be of one colour, juicy, and sweet.

THE VEGETABLE MARROW.

In its best condition the Vegetable Marrow is worthy of a place in any collection. The finest varieties for exhibition are Long White, Moore's Vegetable Cream, and Pen-y-Byd. Seeds should be sown in March in pots filled with light rich soil, and placed in a night temperature of about 60°, potting the plants off singly into four-inch pots as soon as they have shown the rough leaf, and growing them on in the heat till they are well established. They should be gradually hardened off, and be ready for planting out about the end of May. They will grow well in any sheltered place in the full sun, where a pit may be dug two or three feet deep, and filled with heating material of any kind, on which a barrowful of soil is placed. As soon as the heat rises, the plants should be put in, well watered, and slightly shaded till they begin to grow. They must be protected at night if frost threatens after they are planted. As they grow and extend, the shoots should be regulated and stopped beyond each fruit. When the plants are growing freely, liquid manure should be liberally applied, as the most vigorous growth generally gives the finest fruits. In selecting them for exhibition, the fruit should be young and symmetrical, not old and coarse. The points of merit are : (1.) Quality, the flesh fresh, solid, and tender ; (2.) form, globular or oblong, and free from irregularities ; and (3.) size, which should be moderate, neither too small nor too large of the kind. When cut, the flesh should be thick and sweet.

THE SALAD PLANTS.

In the preparation of salads a large variety of plants are employed more or less by connoisseurs, of which the following are the chief :—

Beet.	Corn Salad.	Onion.
Borage.	Cress.	Purslane.
Burnet.	Dandelion.	Radish.
Celeriac.	Endive.	Rampion.
Celery.	Horse-Radish.	Rape.
Chervil.	Lettuce.	Sorrel.
Chickory.	Mint.	Tarragon.
Chives.	Mustard.	Tomato.

These are generally exhibited in collections of a certain number of kinds; and when well grown and set up in fair quantities in a neat style, on a proper tray or stand, they are objects of much interest to the visitors at a horticultural show. For a collection of twelve kinds, the number usually specified in schedules, the following are a good selection :—1. Beet, 2. Celery, 3. Chervil, 4. Common Cress, 5. Endive, 6. Horse-Radish, 7. Lettuce, 8. Mustard, 9. Onion, 10. Radish, 11. Tarragon, 12. Tomato. The cultivation of salads is a comparatively easy matter, but it is too often much neglected, with the result that the plants are poor, tough, juiceless articles, and very unfit for the purpose of making a rich nourishing salad. They mostly all thrive well in a light, rich, friable loam, and delight in heat and moisture, so that many of them are amen-

able to forcing through the winter and spring, at which seasons
a tender and juicy salad is highly appreciated. In many cases
their qualities as salad plants are greatly improved by forcing ;
and when a good collection of crisp, tender, juicy salads is exhibited in the winter or spring, it commands much attention
and merits a good reward. It is not necessary to go into the
details of their individual culture, which is well known to every
good grower, and is described at length in all works on gardening ; but the stimulating influence of rich top-dressings and
frequent doses of liquid manure is nowhere more clearly evident than in the raising of salad plants and roots for exhibition.
As the majority of the plants are used uncooked in making
a salad, they should always be grown in a cleanly condition ;
and when set up for competition, every part should be clean
and perfect, and all in a fresh, tender state, fit in every way
to make a first-rate salad. All herbaceous stems, such as
Burnet, Chevril, Mint, and Tarragon, should be made up into
neat bunches, and a fair quantity shown ; and all roots and
fruits shown in proportionate quantities. The points of merit
are: (1.) All samples well grown ; (2.) quality, fresh, crisp,
tender, and juicy ; and (3.) the selection good, and all exhibited
in suitable proportions.

THE CULINARY AND SWEET HERBS.

This is a class which is seldom seen at horticultural exhibitions, but the importance of many of the articles in the
culinary and household arts fully entitles them to recognition
in the horticultural prize-list when they are exhibited in first-
rate condition. Among those most frequently grown are the
following :—

Angelica.	Dill.	Purslane.
Balm.	Fennel.	Rosemary.
Basil.	Horehound.	Rue.
Borage.	Hyssop.	Sage.
Caraway.	Lavender.	Savory.
Camomile.	Marigold.	Sorrel.
Coriander.	Marjoram.	Thyme.
Costmary	Mint.	Wormwood.

They should always be exhibited in neat bunches, and, in the season, in a fresh green state, just as they have reached the perfect condition for use. A dozen of good kinds for a collection are: 1. Basil, 2. Borage, 3. Fennel, 4. Hyssop, 5. Lavender, 6. Marigold, 7. Marjoram, 8. Mint, 9. Sage, 10. Savory, 11. Sorrel, and 12. Thyme. As "dried herbs," most of them are suitable for exhibition through the winter and spring, and when shown in first-rate condition they form a very interesting exhibit. They are all easily grown in any friable soil in good condition, and when exhibition samples are wanted they are much improved by a little extra attention to manuring, watering in dry weather, and supporting with stakes those that require it. An excellent method of exhibiting a collection of them is to grow them specially for exhibition in pots. Healthy young plants soon form nice specimens in suitable sized pots when grown in an open compost of fibry loam, leaf-mould, and crushed bones. Plunged in a sheltered border, the plants will grow vigorously, and can be fed to any necessary extent with liquid manure and rich surface dressings. A little staking and regulating of those requiring it will make capital specimens of them for exhibition. The points of merit in green or dried herbs are: (1.) Good samples of each in perfect condition; (2.) flavour and quality perfect of their kind; and (3.) kinds well selected and meritorious.

MANURES AND THEIR APPLICATION.

In supplementing what has already been said regarding the soils and manures best adapted to the various plants, we will consider in a general way and briefly the origin and application of plant food, and its effects on the life and growth of the plant.   Chemists tells us that the four primary elements, viz., Carbon, Oxygen, Hydrogen, and Nitrogen form the structure of all plants, and that these simple elements in various combinations and proportions constitute the main body of living organisms as well as of all dead substances in the universe.

Chemical science has of late done much to enlighten the horticulturist on the composition of the soil, the constituents of plants, and the ingredients of manure.   The success or non-success of the cultivator mainly depends upon his bringing into practical operation the principles laid down by the scientist, and the laws which govern supply and demand.

The first thing one naturally inquires about when commencing to cultivate any particular plant or crop is, what manure should be applied so that the best results may be obtained ? If he is conversant with the constitution of the plant, the character of its surroundings, and the nature of its food, he can easily supply its wants.   But many are in want of this knowledge, which is so essential in the economy of the garden or the field, and the lack of which is soon made manifest by the condition of the crops.   The common rule is to apply farmyard manure in greater or less quantities to fertilise soils on which all kinds of crops are to be cultivated.   When this manure is collected from all sources, mixed together, and decomposed, it is supposed to contain within itself in certain proportions all the ingredients of which plants are composed, and which they require as food.   But farmyard manure alone does not contain all the necessary ingredients in the required proportions to promote a strong healthy growth through the various stages and changing conditions of the life of the majority of plants.   Analogy between the mode and means of subsistence in the animal and vegetable kingdoms is generally recognised.   If the animal be supplied, according to its kind, with nutritious digestible food, containing all the elements required for the formation of bone, muscle, and flesh, a good specimen should be the result.   In like manner plants should have within their reach at the proper time the proportions required of the various elements which go to compose their structure, and the formation of flower, fruit, and seed.   Farmyard manure being deficient in inorganic substances, such as potash and lime, should not be given alone to root crops, especially when the mineral elements have been withdrawn from the soil by the frequent repetition of crops of the same nature.   For example, if potatoes were grown for several years in succession in the same place, the crop would gradually become lighter and inferior in quality.   If this system were continued for any length of time, although farmyard manure be abundantly supplied, the ground would cease to bear from lack of sufficient mineral to decompose the organic matter.   By applying lime, which has the power of liberating by chemical action the fertilising properties in vegetable matter, and by returning to the soil the other minerals removed by cropping, fertility is again restored.   On the other hand, if inorganic manures be applied to such crops as brassica and pulse, and the soil be deficient in humus, light crops will follow.   The one class of manure is entirely dependent upon the presence and action of the other for the conversion of both into a suitable condition for absorption and assimilation by plants, and it is evident that if the materials required for the construction, development, and maturation of the plant are not present in sufficient quantities and proportions, the growth will be slow and the parts imperfect. It is, therefore, necessary to apply to the soil the elements in which it is deficient by using artificial manures, most of which are good when properly prepared and unadulterated, and have proved a decided gain to the modern horticulturist.   These manures would be of more value, and could be better utilised by horticulturists, if the ingredients and proportions were generally known. The advantages derived from their use are, first, their easy application either in a dry or a soluble state; second, the choice of ingredients suited to the various crops; third, the assistance they render to farmyard manure by supplying in a concentrated form to rapid-growing plants the organic and inorganic elements in the exact proportions required.

A few common garden plants and the special ingredients they require as food may be given. Peas, beans, and all crops grown for their seeds should have manures rich in silicates and phosphates. Potatoes, turnip, and root crops generally should have lime and potash. Cabbage, spinach, and those vegetables cultivated for their leaves should have sulphates and chlorides. Carbonates and phosphates for fruit trees and bushes. Alkalies are always required along with other manurial ingredients.

These are a few of the more important compounds at the command of the cultivator, the ingredients of which should predominate in all manures applied to the crops or plants mentioned. They should not, however, be substituted entirely for farmyard manure, but should be given along with it, according to the wants of the crop and the deficiency in the soil of the food-supply for that crop. The atmosphere is an inexhaustible source of food-supply for plants, which is received from the various gases and vapours which surround them. This supply cannot be controlled by the horticulturist further than by loosening the surface of the ground, that the nourishing dews may the more readily be absorbed by it, and thus reach the roots of the crop. The loosening the soil is also the means of preventing the rain, which is laden with ammonia and carbonic acid gas, more especially after prolonged drought, from running direct into rents and crevices of the earth, instead of being retained by the humus, and the fertilising properties held fast by it, while the pure water is allowed to pass to the drainage.

The Application of Manures.—Previous to applying farmyard manure, the nature of the soil should be considered. If it be of a close, heavy, and clayey nature, the manure should be applied in a fresh state. If light and sandy, the manure should be well decomposed. In the former case, the soil is rendered more open to the access of air, with a free passage for any surplus of water. In the latter case, the decomposed manure has just an opposite tendency, and it also keeps the roots cool and moist in warm dry weather, while its fertilising elements are ready for immediate absorption. Manure should not be deeply buried, except for such plants as carrot and parsnip, but should be kept near the surface, in the hope that where the elements of nutrition are, there the roots will be also; and there is little doubt that if all the materials required by the plants were placed within convenient reach, less root-pruning would be required amongst bushes and fruit-trees. Besides the solid manure, the liquid drainage from the farmyard is of great value. When it is collected in a tank and allowed to ferment and putrefy, its fertilising properties have an immediate effect on vegetation. It should, however, be well diluted before being applied to growing crops. It may be easily spread over unoccupied ground when in a fresh state, and thus prevent the unavoidable loss of volatile gases in the process of fermentation. Decomposed urine from the byre and stable is an excellent manure, owing to the number of fertilising salts it holds in solution. It should be applied to vegetables and all growing plants with care, and when used in fruit-growing, it should be used only in a weak state when the fruit is swelling and the soil exhausted, because of the tendency which an excess of ammonia and nitrogenous substances have to produce gross spongy growth. The error of using stimulating manures without regard to a proper combination of elements may be illustrated by a common practice in agriculture. The farmer may be observed sowing so many hundredweights of nitrate of soda on an acre of his grass fields, and if the weather is moist, the effect, to all appearance, is wonderful; but if a comparison be made between the stock fed on that part of the field dressed with nitre, and another part which had not received any, the difference will be out of all proportion to that expected, showing clearly that certain manurial substances have the power of increasing the growth of the plant, but cannot impart to it the nutritive elements

P

which sustain the life and growth of the animal. Neither in the process of assimilation can an element suited to one part of the plant be changed to supply any deficiency, or occupy the place of another. Plants in themselves have little or no power to substitute one element for another, to enable them to thrive and grow to perfection. It is the cultivator's part to supply what has not already been provided by Nature for this end. Carbonic acid gas is the principal food of young plants, until the stem is reared and the roots extend, after which a gradual increase in the strength and variety of food should be made, and continued till the leaves develop, when more robust material can be assimilated and appropriated by the different parts of the plant for its own uses, and ultimately to be utilised by man.

Artificial or concentrated manures may be applied to the surface of the soil previous to planting or seed-sowing as a first dressing, or they may be slightly forked in ; but for growing crops they should be watered in, unless when applied in showery weather, when it is better done in the natural way. When it is not convenient to apply these manures to the surface, they should be added to the farm-yard manure, and at once dug in, mixing the whole well with the soil in the operation.

It is, however, in the culture of pot and other plants grown under glass that the benefits derived from the use of concentrated manures are most felt and appreciated, as they require more frequent watering than plants in the open air, and the manures can be applied in smaller quantities and greater variety at short intervals. This is especially applicable to plants in pots, where the quantity of soil is limited, and where they are almost entirely dependent upon an artificial supply. The manures can either be spread upon the surface of the soil in the pot previous to watering, or dissolved in the water before being applied, and in every case the effect should be carefully noted for future guidance. It is by closely observing the effect produced by various kinds of manures upon exotic plants that a correct knowledge of their requirements can be ascertained, varieties of the same species often requiring some deviation in the kind or the strength of the nourishment supplied to produce the best results.

These notes on manures are mainly given for the purpose of drawing closer attention to a wide field of useful knowledge, which has still to be largely utilised. The scientist has done his part in opening it up; it is the province of the practical horticulturist to make the best use of it for the benefit of all.

INDEX.

THE END.

CATALOGUE

OF

MESSRS BLACKWOOD & SONS'

PUBLICATIONS.

PHILOSOPHICAL CLASSICS FOR ENGLISH READERS.

EDITED BY WILLIAM KNIGHT, LL.D.,

Professor of Moral Philosophy in the University of St Andrews.

In crown 8vo Volumes, with Portraits, price 3s. 6d.

Now ready—

DESCARTES, by Professor Mahaffy, Dublin.—BUTLER, by Rev. W. Lucas Collins, M.A.—BERKELEY, by Professor Campbell Fraser, Edinburgh.—FICHTE, by Professor Adamson, Owens College, Manchester.—KANT, by Professor Wallace, Oxford.—HAMILTON, by Professor Veitch, Glasgow.—HEGEL, by Professor Edward Caird, Glasgow.—LEIBNIZ, by J. Theodore Merz.

—VICO, by Professor Flint, Edinburgh—HOBBES, by Professor Croom Robertson, London.—HUME, by the Editor.—SPINOZA, by the Very Rev. Principal Caird, Glasgow.—BACON: Part I. The Life by Professor Nichol, Glasgow.—BACON: Part II. Philosophy, by the same Author.—LOCKE, by Professor Campbell Fraser, Edinburgh.

MILL,   .   *In preparation.*

FOREIGN CLASSICS FOR ENGLISH READERS.

EDITED BY MRS OLIPHANT.

In crown 8vo, 2s. 6d.

Contents of the Series.

DANTE, by the Editor. — VOLTAIRE, by General Sir E. B. Hamley, K.C.B. —PASCAL, by Principal Tulloch.—PETRARCH, by Henry Reeve, C.B.—GOETHE, By A. Hayward, Q.C.—MOLIERE, by the Editor and F. Tarver, M.A.—MONTAIGNE, by Rev. W. L. Collins, M.A.—RABELAIS, by Walter Besant, M.A.—CALDERON, by E. J. Hasell.—SAINT SIMON, by Clifton W. Collins, M.A.—CERVANTES, by the Editor. — CORNEILLE AND RACINE, by Henry M. Trollope. — MADAME DE SÉVIGNÉ, by Miss Thackeray.—LA FONTAINE, AND OTHER FRENCH FABULISTS, by Rev. W. Lucas Collins, M.A.—SCHILLER, by James Sime, M.A., Author of 'Lessing, his Life and Writings.'—TASSO, by E. J. Hasell.—ROUSSEAU, by Henry Grey Graham.—ALFRED DE MUSSET, by C. F. Oliphant.

In preparation.

LEOPARDI. By the Editor.

NOW COMPLETE.

ANCIENT CLASSICS FOR ENGLISH READERS.

EDITED BY THE REV. W. LUCAS COLLINS, M.A.

Complete in 28 Vols. crown 8vo, cloth, price 2s. 6d. each. And may also be had in 14 Volumes, strongly and neatly bound, with calf or vellum back, £3, 10s.

Contents of the Series.

HOMER: THE ILIAD, by the Editor.—HOMER: THE ODYSSEY, by the Editor.—HERODOTUS, by George C. Swayne, M.A.—XENOPHON, by Sir Alexander Grant, Bart., LL.D:—EURIPIDES, by W. B. Donne.—ARISTOPHANES, by the Editor.—PLATO, by Clifton W. Collins, M.A.—LUCIAN, by the Editor.—ÆSCHYLUS, by the Right Rev. the Bishop of Colombo.—SOPHOCLES, by Clifton W. Collins, M.A.—HESIOD AND THEOGNIS, by the Rev. J. Davies, M.A.—GREEK ANTHOLOGY, by Lord Neaves.—VIRGIL, by the Editor.—HORACE, by Sir Theodore Martin, K.C.B.—JUVENAL, by Edward Walford, M.A. — PLAUTUS AND TERENCE, by the Editor.—THE COMMENTARIES OF CÆSAR, by Anthony Trollope.—TACITUS, by W. B. Donne.—CICERO, by the Editor.—PLINY'S LETTERS, by the Rev. Alfred Church, M.A., and the Rev. W. J. Brodribb, M.A.—LIVY, by the Editor.—OVID, by the Rev. A. Church, M.A.—CATULLUS, TIBULLUS, AND PROPERTIUS, by the Rev. Jas. Davies, M.A. — DEMOSTHENES, by the Rev. W. J. Brodribb, M.A.—ARISTOTLE, by Sir Alexander Grant, Bart., LL.D.—THUCYDIDES, by the Editor.—LUCRETIUS, by W. H. Mallock, M.A.—PINDAR, by the Rev. F. D. Morice, M.A.

Saturday Review.—"It is difficult to estimate too highly the value of such a series as this in giving 'English readers' an insight, exact as far as it goes, into those olden times which are so remote, and yet to many of us so close."

CATALOGUE

OF

MESSRS BLACKWOOD & SONS'

PUBLICATIONS.

———◆———

ALISON. History of Europe. By Sir ARCHIBALD ALISON, Bart., D.C.L.

1. From the Commencement of the French Revolution to the Battle of Waterloo.
   LIBRARY EDITION, 14 vols., with Portraits. Demy 8vo, £10, 10s.
   ANOTHER EDITION, in 20 vols. crown 8vo, £6.
   PEOPLE'S EDITION, 13 vols. crown 8vo, £2, 11s.
2. Continuation to the Accession of Louis Napoleon.
   LIBRARY EDITION, 8 vols. 8vo, £6, 7s. 6d.
   PEOPLE'S EDITION, 8 vols. crown 8vo, 34s.
3. Epitome of Alison's History of Europe. Twenty-ninth Thousand, 7s. 6d.
4. Atlas to Alison's History of Europe. By A. Keith Johnston.
   LIBRARY EDITION, demy 4to, £3, 3s.
   PEOPLE'S EDITION, 31s. 6d.

——— Life of John Duke of Marlborough. With some Account of his Contemporaries, and of the War of the Succession. Third Edition. 2 vols. 8vo. Portraits and Maps, 30s.

——— Essays: Historical, Political, and Miscellaneous. 3 vols. demy 8vo, 45s.

ACTA SANCTORUM HIBERNIÆ; Ex Codice Salmanticensi. Nunc primum integre edita opera CAROLI DE SMEDT et JOSEPHI DE BACKER, e Soc. Jesu, Hagiographorum Bollandianorum; Auctore et Sumptus Largiente JOANNE PATRICIO MARCHIONE BOTHAE. In One handsome 4to Volume, bound in half roxburghe, £2, 2s.; in paper wrapper, 31s. 6d.

AIRD. Poetical Works of Thomas Aird. Fifth Edition, with Memoir of the Author by the Rev. JARDINE WALLACE, and Portrait. Crown 8vo, 7s. 6d.

ALLARDYCE. The City of Sunshine. By ALEXANDER ALLARDYCE. Three vols. post 8vo, £1, 5s. 6d.

——— Memoir of the Honourable George Keith Elphinstone, K.B., Viscount Keith of Stonehaven, Marischal, Admiral of the Red. 8vo, with Portrait, Illustrations, and Maps, 21s.

ALMOND. Sermons by a Lay Head-master. By HELY HUTCHINSON ALMOND, M.A. Oxon., Head-master of Loretto School. Crown 8vo, 5s.

ANCIENT CLASSICS FOR ENGLISH READERS. Edited by Rev. W. LUCAS COLLINS, M.A. Price 2s. 6d. each. *For list of Vols., see page 2.*

AYTOUN. Lays of the Scottish Cavaliers, and other Poems. By W. EDMONDSTOUNE AYTOUN, D.C.L., Professor of Rhetoric and Belles-Lettres in the University of Edinburgh. New Edition. Fcap. 8vo, 3s. 6d.
Another Edition, being the Thirtieth. Fcap. 8vo, cloth extra, 7s. 6d.
Cheap Edition. Fcap. 8vo. Illustrated Cover. Price 1s. Cloth, 1s. 3d.

——— An Illustrated Edition of the Lays of the Scottish Cavaliers. From designs by Sir NOEL PATON. Small 4to, in gilt cloth, 21s.

——— Bothwell: a Poem. Third Edition. Fcap. 7s. 6d.

——— Poems and Ballads of Goethe. Translated by Professor AYTOUN and Sir THEODORE MARTIN, K.C.B. Third Edition. Fcap., 6s.

AYTOUN. Bon Gaultier's Book of Ballads. By the SAME. Fifteenth Edition. With Illustrations by Doyle, Leech, and Crowquill. Fcap. 8vo, 5s.

———- The Ballads of Scotland. Edited by Professor AYTOUN. Fourth Edition. 2 vols. fcap. 8vo, 12s.

——— Memoir of William E. Aytoun, D.C.L. By Sir THEODORE MARTIN, K.C.B. With Portrait. Post 8vo, 12s.

BACH. On Musical Education and Vocal Culture. By ALBERT B. BACH. Fourth Edition. 8vo, 7s. 6d.

——— The Principles of Singing. A Practical Guide for Vocalists and Teachers. With Course of Vocal Exercises. Crown 8vo, 6s.

——— The Art of Singing. With Musical Exercises for Young People. Crown 8vo, 3s.

——— The Art Ballad: Loewe and Schubert. With Music Illustrations. With a Portrait of LOEWE. Third Edition. Small 4to. 5s.

BALLADS AND POEMS. By MEMBERS OF THE GLASGOW BALLAD CLUB. Crown 8vo, 7s. 6d

BANNATYNE. Handbook of Republican Institutions in the United States of America. Based upon Federal and State Laws, and other reliable sources of information. By DUGALD J. BANNATYNE, Scotch Solicitor, New York ; Member of the Faculty of Procurators, Glasgow. Cr. 8vo, 7s. 6d.

BELLAIRS. The Transvaal War, 1880-81. Edited by Lady BELLAIRS. With a Frontispiece and Map. 8vo, 15s.

——— Gossips with Girls and Maidens, Betrothed and Free. New Edition. Crown 8vo, 3s. 6d. Cloth, extra gilt edges, 5s.

BESANT. The Revolt of Man. By WALTER BESANT, M.A. Ninth Edition. Crown 8vo, 3s. 6d.

——— Readings in Rabelais. Crown 8vo, 7s. 6d.

BEVERIDGE. Culross and Tulliallan; or Perthshire on Forth. Its History and Antiquities. With Elucidations of Scottish Life and Character from the Burgh and Kirk-Session Records of that District. By DAVID BEVERIDGE. 2 vols. 8vo, with Illustrations, 42s.

——— Between the Ochils and the Forth ; or, From Stirling Bridge to Aberdour. Crown 8vo, 6s.

BLACK. Heligoland and the Islands of the North Sea. By WILLIAM GEORGE BLACK. Crown 8vo, 4s.

BLACKIE. Lays and Legends of Ancient Greece. By JOHN STUART BLACKIE, Emeritus Professor of Greek in the University of Edinburgh. Second Edition. Fcap. 8vo. 5s.

——— The Wisdom of Goethe. Fcap. 8vo. Cloth, extra gilt, 6s.

——— Scottish Song : Its Wealth, Wisdom, and Social Significance. Crown 8vo. With Music. 7s. 6d.

——— A Song of Heroes. Crown 8vo, 6s.

BLACKWOOD'S MAGAZINE, from Commencement in 1817 to April 1892. Nos. 1 to 918, forming 150 Volumes.

——— Index to Blackwood's Magazine. Vols. 1 to 50. 8vo, 15s.

BLACKWOOD. Tales from Blackwood. Price One Shilling each, in Paper Cover. Sold separately at all Railway Bookstalls.
They may also be had bound in cloth, 18s., and in half calf, richly gilt, 30s.
Or 12 volumes in 6, roxburghe, 21s., and half red morocco, 28s.

——— Tales from Blackwood. New Series. Complete in Twenty-four Shilling Parts. Handsomely bound in 12 vols., cloth, 30s. In leather back, roxburghe style, 37s. 6d. In half calf, gilt, 52s. 6d. In half morocco, 55s.

——— Tales from Blackwood. Third Series. Complete in 6 vols. Handsomely bound in cloth, 15s. ; or in 12 vols. 18s. Bound in roxburghe, 21s. Half calf, 25s Half morocco, 28s. Also in 12 parts, price 1s. each.

——— Travel, Adventure, and Sport. From 'Blackwood's Magazine.' Uniform with 'Tales from Blackwood.' In Twelve Parts, each price 1s. Or handsomely bound in 6 vols., 15s. Half calf, 25s.

BLACKWOOD. New Uniform Series of Three-and-Sixpenny Novels
(Copyright). Crown 8vo, cloth. Now ready :—

REATA. By E. D. Gerard.
BEGGAR MY NEIGHBOUR. By the Same.
THE WATERS OF HERCULES. By the Same.
SONS AND DAUGHTERS. By Mrs Oliphant.
FAIR TO SEE. By L. W. M. Lockhart.
THE REVOLT OF MAN. By Walter Besant.
MINE IS THINE. By L. W. M. Lockhart.
ALTIORA PETO. By Laurence Oliphant
DOUBLES AND QUITS. By L.W. M.Lockhart.
LADY BABY. By D. Gerard.

HURRISH. By the Hon. Emily Lawless.
THE BLACKSMITH OF VOE. By Paul
   Cushing.
THE DILEMMA. By the Author of 'The
   Battle of Dorking.'
MY TRIVIAL LIFE AND MISFORTUNE. By
   A Plain Woman.
POOR NELLIE. By the Same.
PICCADILLY. By Laurence Oliphant. With
   Illustrations.

Others in preparation.

——— Standard Novels. Uniform in size and legibly Printed.
Each Novel complete in one volume.

FLORIN SERIES, Illustrated Boards. Or in New Cloth Binding, 2s. 6d.

TOM CRINGLE'S LOG. By Michael Scott.
THE CRUISE OF THE MIDGE. By the Same.
CYRIL THORNTON. By Captain Hamilton.
ANNALS OF THE PARISH. By John Galt.
THE PROVOST, &c. By John Galt.
SIR ANDREW WYLIE. By John Galt.
THE ENTAIL. By John Galt.
MISS MOLLY. By Beatrice May Butt.
REGINALD DALTON. By J. G. Lockhart.

PEN OWEN. By Dean Hook.
ADAM BLAIR. By J. G. Lockhart.
LADY LEE'S WIDOWHOOD. By General
   Sir E. B. Hamley.
SALEM CHAPEL. By Mrs Oliphant.
THE PERPETUAL CURATE. By Mrs Oli-
   phant.
MISS MARJORIBANKS. By Mrs Oliphant.
JOHN : A Love Story. By Mrs Oliphant.

SHILLING SERIES, Illustrated Cover. Or in New Cloth Binding, 1s. 6d.

THE RECTOR, and THE DOCTOR'S FAMILY.
   By Mrs Oliphant.
THE LIFE OF MANSIE WAUCH. By D. M.
   Moir.
PENINSULAR SCENES AND SKETCHES. By
   F. Hardman.

SIR FRIZZLE PUMPKIN, NIGHTS AT MESS
   &c.
THE SUBALTERN.
LIFE IN THE FAR WEST. By G. F. Ruxton.
VALERIUS : A Roman Story. By J. G.
   Lockhart.

BLACKMORE. The Maid of Sker. By R. D. BLACKMORE, Author
of ' Lorna Doone,' &c. New Edition. Crown 8vo, 6s.

BLAIR. History of the Catholic Church of Scotland. From the
Introduction of Christianity to the Present Day. By ALPHONS BELLESHEIM,
D.D., Canon of Aix-la-Chapelle. Translated, with Notes and Additions, by
D. OSWALD HUNTER BLAIR, O.S.B., Monk of Fort Augustus. Complete in
4 vols. demy 8vo, with Maps. Price 12s. 6d. each.

BONNAR. Biographical Sketch of George Meikle Kemp, Architect
of the Scott Monument, Edinburgh. By THOMAS BONNAR, F.S.A. Scot.,
Author of ' The Present Art Revival,' ' The Past of Art in Scotland,' ' Sugges-
tions for the Picturesque of Interiors,' &c. With Three Portraits and numerous
Illustrations. In One Volume, post 8vo.

BOSCOBEL TRACTS. Relating to the Escape of Charles the
Second after the Battle of Worcester, and his subsequent Adventures. Edited
by J. HUGHES, Esq., A.M. A New Edition, with additional Notes and Illustrations, including Communications from the Rev. R. H. BARHAM, Author of
the 'Ingoldsby Legends.' 8vo, with Engravings, 16s.

BROUGHAM. Memoirs of the Life and Times of Henry Lord
Brougham. Written by HIMSELF. 3 vols. 8vo, £2, 8s. The Volumes are sold
separately, price 16s. each.

BROWN. The Forester : A Practical Treatise on the Planting,
Rearing, and General Management of Forest-trees. By JAMES BROWN, LL.D.,
Inspector of and Reporter on Woods and Forests. Fifth Edition, revised and
enlarged. Royal 8vo, with Engravings, 36s.

BROWN. A Manual of Botany, Anatomical and Physiological.
For the Use of Students. By ROBERT BROWN, M.A., Ph.D. Crown 8vo, with
numerous Illustrations, 12s. 6d.

BRUCE. In Clover and Heather. Poems by WALLACE BRUCE.
New and Enlarged Edition. Crown 8vo, 4s. 6d.
A limited number of Copies of the First Edition, on large hand-made paper, 12s. 6d.

BRYDALL. Art in Scotland ; its Origin and Progress. By ROBERT
BRYDALL Master of St George's Art School of Glasgow. 8vo, 12s. 6d.

BUCHAN. Introductory Text-Book of Meteorology. By ALEX-
ANDER BUCHAN, M.A., F.R.S.E., Secretary of the Scottish Meteorological
Society, &c. Crown 8vo, with 8 Coloured Charts and Engravings, 4s. 6d.

BUCHANAN. The Shire Highlands (East Central Africa). By
JOHN BUCHANAN, Planter at Zomba. Crown 8vo, 5s.

BURBIDGE. Domestic Floriculture, Window Gardening, and
Floral Decorations. Being practical directions for the Propagation, Culture,
and Arrangement of Plants and Flowers as Domestic Ornaments. By F. W.
BURBIDGE. Second Edition. Crown 8vo, with numerous Illustrations, 7s. 6d.

—— Cultivated Plants : Their Propagation and Improvement.
Including Natural and Artificial Hybridisation, Raising from Seed, Cuttings,
and Layers, Grafting and Budding, as applied to the Families and Genera in
Cultivation. Crown 8vo, with numerous Illustrations, 12s. 6d.

BURTON. The History of Scotland : From Agricola's Invasion to
the Extinction of the last Jacobite Insurrection. By JOHN HILL BURTON,
D.C.L., Historiographer-Royal for Scotland. New and Enlarged Edition.
8 vols., and Index. Crown 8vo, £3, 3s.

—— History of the British Empire during the Reign of Queen
Anne. In 3 vols. 8vo. 36s.

—— The Scot Abroad. Third Edition. Crown 8vo, 10s. 6d.

—— The Book-Hunter. New Edition. With Portrait. Crown
8vo, 7s. 6d.

BUTE. The Roman Breviary : Reformed by Order of the Holy
Œcumenical Council of Trent; Published by Order of Pope St Pius V.; and
Revised by Clement VIII. and Urban VIII.; together with the Offices since
granted. Translated out of Latin into English by JOHN, Marquess of Bute,
K.T. In 2 vols. crown 8vo, cloth boards, edges uncut. £2, 2s.

—— The Altus of St Columba. With a Prose Paraphrase and
Notes. In paper cover, 2s. 6d.

BUTLER. Pompeii : Descriptive and Picturesque. By W.
BUTLER. Post 8vo, 5s.

BUTT. Miss Molly. By BEATRICE MAY BUTT. Cheap Edition, 2s.

—— Ingelheim. A Novel. 3 vols. crown 8vo, 25s. 6d.

—— Eugenie. Crown 8vo, 6s. 6d.

—— Elizabeth, and Other Sketches. Crown 8vo, 6s.

—— Novels. New and Uniform Edition. Crown 8vo, each 2s. 6d.
Delicia. Now ready.

CAIRD. Sermons. By JOHN CAIRD, D.D., Principal of the University of Glasgow. Sixteenth Thousand. Fcap. 8vo, 5s.

—— Religion in Common Life. A Sermon preached in Crathie
Church, October 14, 1855, before Her Majesty the Queen and Prince Albert.
Published by Her Majesty's Command. Cheap Edition, 3d.

CALDER. Chaucer's Canterbury Pilgrimage. Epitomised by
WILLIAM CALDER. With Photogravure of the Pilgrimage Company, and other
Illustrations, Glossary, &c. Crown 8vo, 4s.

CAMPBELL. Critical Studies in St Luke's Gospel : Its Demonology
and Ebionitism. By COLIN CAMPBELL, B.D., Minister of the Parish of Dun-
dee, formerly Scholar and Fellow of Glasgow University. Author of the 'Three
First Gospels in Greek, arranged in parallel columns. Post 8vo, 7s. 6d.

CAMPBELL. Sermons Preached before the Queen at Balmoral.
By the Rev. A. A. CAMPBELL, Minister of Crathie. Published by Command
of Her Majesty. Crown 8vo, 4s. 6d.

CAMPBELL. Records of Argyll. Legends, Traditions, and Recollections of Argyllshire Highlanders, collected chiefly from the Gaelic.
With Notes on the Antiquity of the Dress, Clan Colours or Tartans of the
Highlanders. By LORD ARCHIBALD CAMPBELL. Illustrated with Nineteen
full-page Etchings. 4to, printed on hand-made paper, £3, 3s.

CANTON. A Lost Epic, and other Poems. By WILLIAM CANTON.
Crown 8vo, 5s.

CARRICK. Koumiss ; or, Fermented Mare's Milk : and its Uses in the Treatment and Cure of Pulmonary Consumption, and other Wasting Diseases. With an Appendix on the best Methods of Fermenting Cow's Milk. By GEORGE L. CARRICK, M.D., L.R.C.S.E. and L.R.C.P.E., Physician to the British Embassy, St Petersburg, &c. Crown 8vo, 10s. 6d.

CARSTAIRS. British Work in India. By R. CARSTAIRS. Cr. 8vo, 6s.

CAUVIN. A Treasury of the English and German Languages. Compiled from the best Authors and Lexicographers in both Languages. By JOSEPH CAUVIN, LL.D. and Ph.D., of the University of Göttingen, &c. Crown 8vo, 7s. 6d.

CAVE-BROWN. Lambeth Palace and its Associations. By J. CAVE-BROWN, M.A., Vicar of Detling, Kent, and for many years Curate of Lambeth Parish Church. With an Introduction by the Archbishop of Canterbury. Second Edition, containing an additional Chapter on Medieval Life in the Old Palaces. 8vo, with Illustrations, 21s.

CHARTERIS. Canonicity ; or, Early Testimonies to the Existence and Use of the Books of the New Testament. Based on Kirchhoffer's 'Quellensammlung.' Edited by A. H. CHARTERIS, D.D., Professor of Biblical Criticism in the University of Edinburgh. 8vo, 18s.

CHRISTISON. Life of Sir Robert Christison, Bart., M.D., D.C.L. Oxon., Professor of Medical Jurisprudence in the University of Edinburgh. Edited by his SONS. In two vols. 8vo. Vol. I.—Autobiography. 16s. Vol. II. —Memoirs. 16s.

CHRONICLES OF WESTERLY : A Provincial Sketch. By the Author of 'Culmshire Folk,' 'John Orlebar,' &c. 3 vols. crown 8vo, 25s. 6d.

CHURCH SERVICE SOCIETY. A Book of Common Order : Being Forms of Worship issued by the Church Service Society. Sixth Edition. Crown, 8vo, 6s. Also in 2 vols, crown 8vo, 6s. 6d.

CLOUSTON. Popular Tales and Fictions : their Migrations and Transformations. By W. A. CLOUSTON, Editor of 'Arabian Poetry for English Readers,' &c. 2 vols. post 8vo, roxburghe binding, 25s.

COCHRAN. A Handy Text-Book of Military Law. Compiled chiefly to assist Officers preparing for Examination ; also for all Officers of the Regular and Auxiliary Forces. Comprising also a Synopsis of part of the Army Act. By Major F. COCHRAN, Hampshire Regiment Garrison Instructor, North British District. Crown 8vo, 7s. 6d.

COLQUHOUN. The Moor and the Loch. Containing Minute Instructions in all Highland Sports, with Wanderings over Crag and Corrie, Flood and Fell. By JOHN COLQUHOUN. Seventh Edition. With Illustrations. 8vo, 21s.

COTTERILL. Suggested Reforms in Public Schools. By C. C. COTTERILL, M.A. Crown 8vo, 3s. 6d.

CRANSTOUN. The Elegies of Albius Tibullus. Translated into English Verse, with Life of the Poet, and Illustrative Notes. By JAMES CRANSTOUN, LL.D., Author of a Translation of 'Catullus.' Crown 8vo, 6s. 6d.

—————— The Elegies of Sextus Propertius. Translated into English Verse, with Life of the Poet, and Illustrative Notes. Crown 8vo, 7s. 6d.

CRAWFORD. Saracinesca. By F. MARION CRAWFORD, Author of 'Mr Isaacs,' 'Dr Claudius,' 'Zoroaster,' &c. &c. Fifth Ed. Crown 8vo, 6s.

CRAWFORD. The Doctrine of Holy Scripture respecting the Atonement. By the late THOMAS J. CRAWFORD, D.D., Professor of Divinity in the University of Edinburgh. Fifth Edition. 8vo, 12s.

—————— The Fatherhood of God, Considered in its General and Special Aspects. Third Edition, Revised and Enlarged. 8vo, 9s.

—————— The Preaching of the Cross, and other Sermons. 8vo, 7s. 6d.

—————— The Mysteries of Christianity. Crown 8vo, 7s. 6d.

CRAWFORD. An Atonement of East London, and other Poems. By HOWARD CRAWFORD, M.A. Crown 8vo, 5s.

CUSHING. The Blacksmith of Voe. By PAUL CUSHING, Author of 'The Bull i' th' Thorn.' Cheap Edition. Crown 8vo, 3s. 6d.

—————— Cut with his own Diamond. A Novel. 3 vols. cr. 8vo, 25s. 6d.

DAVIES.   Norfolk Broads and Rivers ; or, The Waterways, Lagoons,
and Decoys of East Anglia. By G. CHRISTOPHER DAVIES. Illustrated with
Seven full-page Plates. New and Cheaper Edition. Crown 8vo, 6s.

——— Our Home in Aveyron.   Sketches of Peasant Life in
Aveyron and the Lot.   By G. CHRISTOPHER DAVIES and Mrs BROUGHALL.
Illustrated with full-page Illustrations. 8vo, 15s. Cheap Edition, 7s. 6d.

DAYNE.   Tribute to Satan.   A Novel.   By J. BELFORD DAYNE,
Author of 'In the Name of the Tzar.' Crown 8vo, 2s. 6d.

DE LA WARR.   An Eastern Cruise in the 'Edeline.'   By the
Countess DE LA WARR.   In Illustrated Cover.   2s.

DESCARTES.   The Method, Meditations, and Principles of Philo-
sophy of Descartes. Translated from the Original French and Latin. With a
New Introductory Essay, Historical and Critical, on the Cartesian Philosophy.
By Professor VEITCH, LL.D., Glasgow University. Ninth Edition. 6s. 6d.

DICKSON.   Gleanings from Japan.   By W. G. DICKSON, Author
of 'Japan: Being a Sketch of its History, Government, and Officers of the
Empire.' With Illustrations. 8vo, 16s.

DOGS, OUR DOMESTICATED : Their Treatment in reference
to Food, Diseases, Habits, Punishment, Accomplishments. By 'MAGENTA.'
Crown 8vo, 2s. 6d.

DOMESTIC EXPERIMENT, A.   By the Author of 'Ideala : A
Study from Life.' Crown 8vo, 6s.

DR HERMIONE.   By the Author of 'Lady Bluebeard,' 'Zit and
Xoe.' Crown 8vo, 6s.

DU CANE.   The Odyssey of Homer, Books I.-XII.   Translated into
English Verse. By Sir CHARLES DU CANE, K.C.M.G. 8vo, 10s. 6d.

DUDGEON.   History of the Edinburgh or Queen's Regiment
Light Infantry Militia, now 3rd Battalion The Royal Scots; with an
Account of the Origin and Progress of the Militia, and a Brief Sketch of the
old Royal Scots. By Major R. C. DUDGEON, Adjutant 3rd Battalion The Royal
Scots. Post 8vo, with Illustrations, 10s. 6d.

DUNCAN.   Manual of the General Acts of Parliament relating to
the Salmon Fisheries of Scotland from 1828 to 1882. By J. BARKER DUNCAN.
Crown 8vo, 5s.

DUNSMORE.   Manual of the Law of Scotland as to the Relations
between Agricultural Tenants and their Landlords, Servants, Merchants, and
Bowers. By W. DUNSMORE. 8vo, 7s. 6d.

DUPRÉ.   Thoughts on Art, and Autobiographical Memoirs of
Giovanni Duprè. Translated from the Italian by E. M. PERUZZI, with the
permission of the Author. New Edition. With an Introduction by W. W.
STORY. Crown 8vo, 10s. 6d.

ELIOT.   George Eliot's Life, Related in her Letters and Journals.
Arranged and Edited by her husband, J. W. CROSS. With Portrait and other
Illustrations. Third Edition. 3 vols. post 8vo, 42s.

——— George Eliot's Life.   (Cabinet Edition.) With Portrait
and other Illustrations. 3 vols. crown 8vo, 15s.

——— George Eliot's Life.   With Portrait and other Illustrations.
New Edition, in one volume. Crown 8vo, 7s. 6d.

——— Works of George Eliot (Cabinet Edition).   Handsomely
printed in a new type, 21 volumes, crown 8vo, price £5, 5s. The Volumes
are also sold separately, price 5s. each, viz. :—
Romola. 2 vols.—Silas Marner, The Lifted Veil, Brother Jacob. 1 vol.—
Adam Bede. 2 vols.—Scenes of Clerical Life. 2 vols.—The Mill on the Floss.
2 vols.—Felix Holt. 2 vols.—Middlemarch. 3 vols.—Daniel Deronda. 3
vols.—The Spanish Gypsy. 1 vol.—Jubal, and other Poems, Old and New.
1 vol.—Theophrastus Such. 1 vol.—Essays. 1 vol.

——— Novels by GEORGE ELIOT.   Cheap Edition.   Adam Bede. Illustrated. 3s. 6d., cloth.—The Mill on the Floss. Illustrated. 3s. 6d., cloth.
—Scenes of Clerical Life. Illustrated. 3s., cloth.—Silas Marner : the Weaver
of Raveloe. Illustrated. 2s. 6d., cloth.—Felix Holt, the Radical. Illustrated. 3s. 6d., cloth.—Romola. With Vignette. 3s. 6d., cloth.

——— Middlemarch.   Crown 8vo, 7s. 6d.

ELIOT. Daniel Deronda. Crown 8vo, 7s. 6d.
——— Essays. New Edition. Crown 8vo, 5s.
——— Impressions of Theophrastus Such. New Ed. Cr. 8vo, 5s.
——— The Spanish Gypsy. New Edition. Crown 8vo, 5s.
——— The Legend of Jubal, and other Poems, Old and New.
New Edition. Crown 8vo, 5s.
——— Wise, Witty, and Tender Sayings, in Prose and Verse.
Selected from the Works of GEORGE ELIOT. Eighth Edition. Fcap. 8vo, 6s.
——— The George Eliot Birthday Book. Printed on fine paper,
with red border, and handsomely bound in cloth, gilt. Fcap. 8vo, cloth, 3s. 6d.
And in French morocco or Russia, 5s.
ESSAYS ON SOCIAL SUBJECTS. Originally published in the
'Saturday Review.' New Ed. First & Second Series. 2 vols. cr. 8vo, 6s. each.
EWALD. The Crown and its Advisers ; or, Queen, Ministers,
Lords and Commons. By ALEXANDER CHARLES EWALD, F.S.A. Crown 8vo, 5s.
FAITHS OF THE WORLD, The. A Concise History of the
Great Religious Systems of the World. By various Authors. Crown 8vo, 5s.
FARRER. A Tour in Greece in 1880. By RICHARD RIDLEY
FARRER. With Twenty-seven full-page Illustrations by LORD WINDSOR.
Royal 8vo, with a Map, 21s.
FERRIER. Philosophical Works of the late James F. Ferrier,
B.A. Oxon., Professor of Moral Philosophy and Political Economy, St Andrews.
New Edition. Edited by Sir ALEX. GRANT, Bart., D.C.L., and Professor
LUSHINGTON. 3 vols. crown 8vo, 34s. 6d.
——— Institutes of Metaphysic. Third Edition. 10s. 6d.
——— Lectures on the Early Greek Philosophy. 3d Ed. 10s. 6d.
——— Philosophical Remains, including the Lectures on Early
Greek Philosophy. 2 vols., 24s.
FITZROY. Dogma and the Church of England. By A. I. FITZROY.
Post 8vo, 7s. 6d.
FLINT. The Philosophy of History in Europe. By ROBERT
FLINT, D.D., LL.D., Professor of Divinity, University of Edinburgh. 2 vols.
8vo. [New Edition in preparation.
——— Theism. Being the Baird Lecture for 1876. Eighth Edi-
tion. Crown 8vo, 7s. 6d.
——— Anti-Theistic Theories. Being the Baird Lecture for 1877.
Fourth Edition. Crown 8vo, 10s. 6d.
——— Agnosticism. Being the Croall Lectures for 1887-88.
[In the press.
FORBES. Insulinde : Experiences of a Naturalist's Wife in the
Eastern Archipelago. By Mrs H. O. FORBES. Crown 8vo, with a Map. 4s. 6d.
FOREIGN CLASSICS FOR ENGLISH READERS. Edited
by Mrs OLIPHANT. Price 2s. 6d. For List of Volumes published, see page 2.
FOSTER. The Fallen City, and Other Poems. By WILL FOSTER.
Crown 8vo, 6s.
FRANCILLON. Gods and Heroes ; or, The Kingdom of Jupiter.
By R. E. FRANCILLON. With 8 Illustrations. Crown 8vo, 5s.
FULLARTON. Merlin : A Dramatic Poem. By RALPH MACLEOD
FULLARTON. Crown 8vo, 5s.
GALT. Novels by JOHN GALT. Fcap. 8vo, boards, 2s.; cloth, 2s. 6d.
Annals of the Parish.—The Provost.—Sir Andrew Wylie.—
The Entail.
GENERAL ASSEMBLY OF THE CHURCH OF SCOTLAND.
——— Prayers for Social and Family Worship. Prepared by a
Special Committee of the General Assembly of the Church of Scotland. En-
tirely New Edition, Revised and Enlarged. Fcap. 8vo, red edges, 2s.
——— Prayers for Family Worship. A Selection from the com-
plete book. Fcap. 8vo, red edges, price 1s.

GENERAL ASSEMBLY OF THE CHURCH OF SCOTLAND.
———— Scottish Hymnal, with Appendix Incorporated. Published for Use in Churches by Authority of the General Assembly. 1. Large type, cloth, red edges, 2s. 6d. ; French morocco, 4s. 2. Bourgeois type, limp cloth, 1s.; French morocco, 2s. 3. Nonpareil type, cloth, red edges, 6d. ; French morocco, 1s. 4d. 4. Paper covers, 3d. 5. Sunday-School Edition, paper covers, 1d. No. 1, bound with the Psalms and Paraphrases, French morocco, 8s. No. 2, bound with the Psalms and Paraphrases, cloth, 2s. ; French morocco, 3s.

GERARD. Reata: What's in a Name. By E. D. GERARD. Cheap Edition. Crown 8vo, 3s. 6d.

———— Beggar my Neighbour. Cheap Edition. Crown 8vo, 3s. 6d.

———— The Waters of Hercules. Cheap Edition. Crown 8vo, 3s. 6d.

GERARD. The Land beyond the Forest. Facts, Figures, and Fancies from Transylvania. By E. GERARD. In Two Volumes. With Maps and Illustrations. 25s.

———— Bis: Some Tales Retold. Crown 8vo, 6s.

———— A Secret Mission. 2 vols. crown 8vo, 17s.

GERARD. Lady Baby. By DOROTHEA GERARD, Author of 'Orthodox.' Cheap Edition. Crown 8vo, 3s. 6d.

———— Recha. Second Edition. Crown 8vo, 6s.

GERARD. Stonyhurst Latin Grammar. By Rev. JOHN GERARD. [New Edition in preparation.

GILL. Free Trade: an Inquiry into the Nature of its Operation. By RICHARD GILL. Crown 8vo, 7s. 6d.

———— Free Trade under Protection. Crown 8vo, 7s. 6d.

GOETHE'S FAUST. Translated into English Verse by Sir THEODORE MARTIN, K.C.B. Part I. Second Edition, post 8vo, 6s. Ninth Edition, fcap., 3s. 6d. Part II. Second Edition, revised. Fcap. 8vo, 6s.

GOETHE. Poems and Ballads of Goethe. Translated by Professor AYTOUN and Sir THEODORE MARTIN, K.C.B. Third Edition, fcap. 8vo, 6s.

GOODALL. Juxta Crucem. Studies of the Love that is over us. By the late Rev. CHARLES GOODALL, B.D., Minister of Barr. With a Memoir by Rev. Dr Strong, Glasgow, and Portrait. Crown 8vo, 6s.

GORDON CUMMING. Two Happy Years in Ceylon. By C. F. GORDON CUMMING. With 15 full-page Illustrations and a Map. Fourth Edition. 2 vols. 8vo, 30s.

———— At Home in Fiji. Fourth Edition, post 8vo. With Illustrations and Map. 7s. 6d.

———— A Lady's Cruise in a French Man-of-War. New and Cheaper Edition. 8vo. With Illustrations and Map. 12s. 6d.

———— Fire-Fountains. The Kingdom of Hawaii: Its Volcanoes, and the History of its Missions. With Map and Illustrations. 2 vols. 8vo, 25s.

———— Wanderings in China. New and Cheaper Edition. 8vo, with Illustrations, 10s.

———— Granite Crags: The Yō-semité Region of California. Illustrated with 8 Engravings. New and Cheaper Edition. 8vo, 8s. 6d

GRAHAM. The Life and Work of Syed Ahmed Khan, C.S.I. By Lieut.-Colonel G. F. I. GRAHAM, B.S.C. 8vo, 14s

GRAHAM. Manual of the Elections (Scot.) (Corrupt and Illegal Practices) Act, 1890. With Analysis, Relative Act of Sederunt, Appendix containing the Corrupt Practices Acts of 1883 and 1885, and Copious Index. By J. EDWARD GRAHAM, Advocate. 8vo, 4s. 6d.

GRANT. Bush-Life in Queensland. By A. C. GRANT. New Edition. Crown 8vo, 6s.

GRIFFITHS. Locked Up. By Major ARTHUR GRIFFITHS, Author of 'The Wrong Road,' 'Chronicles of Newgate,' &c. With Illustrations by C. J. STANILAND, R.I. Crown 8vo, 2s. 6d.

GUTHRIE-SMITH. Crispus: A Drama. By H. GUTHRIE-SMITH. In one volume. Fcap. 4to, 5s.

HAINES. Unless! A Romance. By RANDOLPH HAINES. Crown 8vo, 6s.

HALDANE. Subtropical Cultivations and Climates. A Handy Book for Planters, Colonists, and Settlers. By R. C. HALDANE. Post 8vo, 9s.

HALLETT. A Thousand Miles on an Elephant in the Shan States. By HOLT S. HALLETT, M. Inst. C.E., F.R.G.S., M.R.A.S., Hon. Member Manchester and Tyneside Geographical Societies. 8vo, with Maps and numerous Illustrations, 21s.

HAMERTON. Wenderholme : A Story of Lancashire and Yorkshire Life. By PHILIP GILBERT HAMERTON, Author of 'A Painter's Camp.' A New Edition. Crown 8vo, 6s.

HAMILTON. Lectures on Metaphysics. By Sir WILLIAM HAMILTON, Bart., Professor of Logic and Metaphysics in the University of Edinburgh. Edited by the Rev. H. L. MANSEL, B.D., LL.D., Dean of St Paul's ; and JOHN VEITCH, M.A., LL.D., Professor of Logic and Rhetoric, Glasgow. Seventh Edition. 2 vols. 8vo, 24s.

—— Lectures on Logic. Edited by the SAME. Third Edition. 2 vols., 24s.

—— Discussions on Philosophy and Literature, Education and University Reform. Third Edition, 8vo, 21s.

—— Memoir of Sir William Hamilton, Bart., Professor of Logic and Metaphysics in the University of Edinburgh. By Professor VEITCH, of the University of Glasgow. 8vo, with Portrait, 18s.

—— Sir William Hamilton : The Man and his Philosophy. Two Lectures delivered before the Edinburgh Philosophical Institution, January and February 1883. By the SAME. Crown 8vo, 2s.

HAMLEY. The Operations of War Explained and Illustrated. By General Sir EDWARD BRUCE HAMLEY, K.C.B., K.C.M.G., M.P. Fifth Edition, revised throughout. 4to, with numerous Illustrations, 30s.

—— National Defence ; Articles and Speeches. Post 8vo, 6s.

—— Shakespeare's Funeral, and other Papers. Post 8vo, 7s. 6d.

—— Thomas Carlyle : An Essay. Second Ed. Cr. 8vo, 2s. 6d.

—— On Outposts. Second Edition. 8vo, 2s.

—— Wellington's Career ; A Military and Political Summary. Crown 8vo, 2s.

—— Lady Lee's Widowhood. Crown 8vo, 2s. 6d.

—— Our Poor Relations. A Philozoic Essay. With Illustrations, chiefly by Ernest Griset Crown 8vo, cloth gilt, 3s. 6d.

HAMLEY. Guilty, or Not Guilty? A Tale. By Major-General W. G. HAMLEY, late of the Royal Engineers. New Edition. Crown 8vo, 3s. 6d.

HARRISON. The Scot in Ulster. The Story of the Scottish Settlement in Ulster. By JOHN HARRISON, Author of ' Oure Tounis College.' Crown 8vo, 2s. 6d.

HASELL. Bible Partings. By E. J. HASELL. Crown 8vo, 6s.

—— Short Family Prayers. Cloth, 1s.

HAY. The Works of the Right Rev. Dr George Hay, Bishop of Edinburgh. Edited under the Supervision of the Right Rev. Bishop STRAIN. With Memoir and Portrait of the Author. 5 vols. crown 8vo, bound in extra cloth, £1, 1s. The following Volumes may be had separately—viz. : The Devout Christian Instructed in the Law of Christ from the Written Word. 2 vols., 8s.—The Pious Christian Instructed in the Nature and Practice of the Principal Exercises of Piety. 1 vol., 3s.

HEATLEY. The Horse-Owner's Safeguard. A Handy Medical Guide for every Man who owns a Horse. By G. S. HEATLEY, M.R.C.V.S. Crown 8vo, 5s.

—— The Stock-Owner's Guide. A Handy Medical Treatise for every Man who owns an Ox or a Cow. Crown 8vo, 4s. 6d.

HEDDERWICK. Lays of Middle Age ; and other Poems. By JAMES HEDDERWICK, LL.D. Price 3s. 6d.

HEDDERWICK.   Backward Glances; or, Some Personal Recollections.   With a Portrait.   Post 8vo, 7s. 6d.

HEMANS.   The Poetical Works of Mrs Hemans.   Copyright Edi-
tions.—Royal 8vo, 5s.—The Same, with Engravings, cloth, gilt edges, 7s. 6d.
—Six Vols. in Three, fcap., 12s. 6d.
SELECT POEMS OF MRS HEMANS.   Fcap., cloth, gilt edges, 3s.

HERKLESS.   Cardinal Beaton   Priest and Politician.   By JOHN
HERKLESS, Minister of Tannadice.   With a Portrait.   Post 8vo, 7s. 6d.

HOME PRAYERS.   By Ministers of the Church of Scotland and
Members of the Church Service Society.   Second Edition.   Fcap. 8vo, 3s.

HOMER.   The Odyssey.   Translated into English Verse in the
Spenserian Stanza.   By PHILIP STANHOPE WORSLEY.   Third Edition, 2 vols.
fcap., 12s.

——— The Iliad.   Translated by P. S. WORSLEY and Professor
CONINGTON.   2 vols. crown 8vo, 21s.

HUTCHINSON.   Hints on the Game of Golf.   By HORACE G.
HUTCHINSON.   Sixth Edition, Enlarged.   Fcap. 8vo, cloth, 1s.

IDDESLEIGH.   Lectures and Essays.   By the late EARL OF
IDDESLEIGH, G.C.B., D.C.L., &c.   8vo, 16s.

——— Life, Letters, and Diaries of Sir Stafford Northcote, First
Earl of Iddesleigh.   By ANDREW LANG.   With Three Portraits and a View of
Pynes.   Third Edition.   2 vols   Post 8vo, 31s. 6d.
POPULAR EDITION.   In one volume.   With two Plates.   Post 8vo, 7s. 6d.

INDEX GEOGRAPHICUS : Being a List, alphabetically arranged,
of the Principal Places on the Globe, with the Countries and Subdivisions of
the Countries in which they are situated, and their Latitudes and Longitudes.
Imperial 8vo, pp. 676, 21s.

JEAN JAMBON.   Our Trip to Blunderland ; or, Grand Excursion
to Blundertown and Back.   By JEAN JAMBON.   With Sixty Illustrations
designed by CHARLES DOYLE, engraved by DALZIEL.   Fourth Thousand.
Cloth, gilt edges, 6s. 6d.   Cheap Edition, cloth, 3s. 6d.   Boards, 2s. 6d.

JENNINGS.   Mr Gladstone : A Study.   By LOUIS J. JENNINGS,
M.P., Author of 'Republican Government in the United States,' 'The Croker
Memoirs,' &c.   Popular Edition.   Crown 8vo, 1s.

JERNINGHAM.   Reminiscences of an Attaché.   By HUBERT
E. H. JERNINGHAM.   Second Edition.   Crown 8vo, 5s.

——— Diane de Breteuille.   A Love Story.   Crown 8vo, 2s. 6d.

JOHNSTON.   The Chemistry of Common Life.   By Professor
J. F. W. JOHNSTON.   New Edition, Revised, and brought down to date.   By
ARTHUR HERBERT CHURCH, M.A. Oxon.; Author of 'Food: its Sources,
Constituents, and Uses,' &c.   With Maps and 102 Engravings.   Cr. 8vo, 7s. 6d.

——— Elements of Agricultural Chemistry and Geology.   Revised, and brought down to date.   By Sir CHARLES A. CAMERON, M.D.,
F.R.C.S.I., &c.   Sixteenth Edition.   Fcap. 8vo, 6s. 6d.

——— Catechism of Agricultural Chemistry and Geology.   Revised by Sir C. A. CAMERON.   With numerous Illustrations.
[New Edition in preparation.

JOHNSTON.   Patrick Hamilton : a Tragedy of the Reformation
in Scotland, 1528.   By T P. JOHNSTON.   Crown 8vo, with Two Etchings.   5s.

KEBBEL.   The Old and the New : English Country Life.   The
Country Clergy—The Country Gentlemen—The Farmers—The Peasantry—
The Eighteenth Century.   By T. E. KEBBEL, M A, Author of 'Agricultural
Labourers,' 'Essays in History and Politics,' 'Life of Lord Beaconsfield.'
Crown 8vo, 5s.

KING.   The Metamorphoses of Ovid.   Translated in English Blank
Verse.   By HENRY KING, M.A., Fellow of Wadham College, Oxford, and of
the Inner Temple, Barrister-at-Law.   Crown 8vo, 10s. 6d.

KINGLAKE.   History of the Invasion of the Crimea.   By A. W.
KINGLAKE.   Cabinet Edition, revised.   With an Index to the Complete Work.
Illustrated with Maps and Plans.   Complete in 9 Vols., crown 8vo, at 6s. each.

KINGLAKE. History of the Invasion of the Crimea. Demy 8vo.
Vol. VI. Winter Troubles. With a Map, 16s. Vols. VII. and VIII. From
the Morrow of Inkerman to the Death of Lord Raglan. With an Index to
the Whole Work. With Maps and Plans. 28s.

———— Eothen. A New Edition, uniform with the Cabinet Edition
of the 'History of the Invasion of the Crimea,' price 6s.

KNEIPP. My Water-Cure. As Tested through more than Thirty
Years, and Described for the Healing of Diseases and the Preservation of Health.
By SEBASTIAN KNEIPP, Parish Priest of Wörishofen (Bavaria). With a Portrait
and other Illustrations. Only Authorised English Translation. Translated from
the Thirtieth German Edition by A. de F. Crown 8vo, 5s.

KNOLLYS. The Elements of Field-Artillery. Designed for the
Use of Infantry and Cavalry Officers. By HENRY KNOLLYS, Captain Royal
Artillery; Author of 'From Sedan to Saarbrück,' Editor of 'Incidents in the
Sepoy War,' &c. With Engravings. Crown 8vo, 7s. 6d.

LAMINGTON. In the Days of the Dandies. By the late Lord
LAMINGTON. Crown 8vo. Illustrated cover, 1s.; cloth, 1s. 6d.

LAWLESS. Hurrish: a Study. By the Hon. EMILY LAWLESS,
Author of 'A Chelsea Householder,' &c. Fourth Edition, crown 8vo, 3s. 6d.

LAWSON. Spain of To-day: A Descriptive, Industrial, and Finan-
cial Survey of the Peninsula, with a full account of the Rio Tinto Mines. By
W. R. LAWSON. Crown 8vo, 3s 6d.

LEES. A Handbook of the Sheriff and Justice of Peace Small
Debt Courts. 8vo, 7s. 6d.

LIGHTFOOT. Studies in Philosophy. By the Rev. J. LIGHTFOOT,
M.A., D.Sc., Vicar of Cross Stone, Todmorden. Crown 8vo, 4s. 6d.

LLOYD. Ireland under the Land League. A Narrative of Personal
Experiences. By CLIFFORD LLOYD, Special Resident Magistrate. Post 8vo, 6s.

LOCKHART. Novels by LAURENCE W. M. LOCKHART. See
Blackwoods' New Series of Three-and-Sixpenny Novels on page 5.

LORIMER. The Institutes of Law: A Treatise of the Principles
of Jurisprudence as determined by Nature. By the late JAMES LORIMER,
Professor of Public Law and of the Law of Nature and Nations in the University of Edinburgh. New Edition, revised and much enlarged. 8vo, 18s.

———— The Institutes of the Law of Nations. A Treatise of the
Jural Relation of Separate Political Communities. In 2 vols. 8vo. Volume
I., price 16s. Volume II., price 20s.

LOVE. Scottish Church Music. Its Composers and Sources. With
Musical Illustrations. By JAMES LOVE. In 1 vol. post 8vo, 7s. 6d.

M'COMBIE. Cattle and Cattle-Breeders. By WILLIAM M'COMBIE,
Tillyfour. New Edition, enlarged, with Memoir of the Author. By JAMES
MACDONALD, of the 'Farming World.' Crown 8vo, 3s. 6d.

MACRAE. A Handbook of Deer-Stalking. By ALEXANDER
MACRAE, late Forester to Lord Henry Bentinck. With Introduction by
HORATIO ROSS, Esq. Fcap. 8vo, with two Photographs from Life. 3s. 6d.

M'CRIE. Works of the Rev. Thomas M'Crie, D.D. Uniform Edition. Four vols. crown 8vo, 24s.

———— Life of John Knox. Containing Illustrations of the History of the Reformation in Scotland. Crown 8vo, 6s. Another Edition, 3s. 6d.

———— Life of Andrew Melville. Containing Illustrations of the
Ecclesiastical and Literary History of Scotland in the Sixteenth and Seventeenth Centuries. Crown 8vo, 6s.

———— History of the Progress and Suppression of the Reformation in Italy in the Sixteenth Century. Crown 8vo, 4s.

———— History of the Progress and Suppression of the Reformation in Spain in the Sixteenth Century. Crown 8vo, 3s. 6d.

———— Lectures on the Book of Esther. Fcap. 8vo, 5s.

MACDONALD. A Manual of the Criminal Law (Scotland) Procedure Act, 1887. By NORMAN DORAN MACDONALD. Revised by the LORD
JUSTICE-CLERK. 8vo, cloth, 10s. 6d.

14 LIST OF BOOKS PUBLISHED BY

MACGREGOR. Life and Opinions of Major-General Sir Charles MacGregor, K.C.B., C.S.I., C.I.E, Quartermaster-General of India. From his Letters and Diaries. Edited by LADY MACGREGOR. With Portraits and Maps to illustrate Campaigns in which he was engaged. 2 vols. 8vo, 35s.

M'INTOSH. The Book of the Garden. By CHARLES M'INTOSH, formerly Curator of the Royal Gardens of his Majesty the King of the Belgians, and lately of those of his Grace the Duke of Buccleuch, K.G., at Dalkeith Palace. 2 vols. royal 8vo, with 1350 Engravings. £4, 7s. 6d. Vol. I. On the Formation of Gardens and Construction of Garden Edifices. £2, 10s. Vol. II. Practical Gardening. £1, 17s. 6d.

MACINTYRE. Hindu-Koh: Wanderings and Wild Sports on and beyond the Himalayas. By Major-General DONALD MACINTYRE, V.C., late Prince of Wales' Own Goorkhas, F.R.G.S. *Dedicated to H.R.H. The Prince of Wales.* New and Cheaper Edition, revised, with numerous Illustrations, post 8vo, 7s. 6d.

MACKAY. A Sketch of the History of Fife and Kinross. A Study of Scottish History and Character. By Æ. J. G. MACKAY, Sheriff of these Counties. Crown 8vo, 6s.

MACKAY. A Manual of Modern Geography; Mathematical, Physical, and Political. By the Rev. ALEXANDER MACKAY, LL.D., F.R.G.S. 11th Thousand, revised to the present time. Crown 8vo, pp. 688. 7s. 6d.

—— Elements of Modern Geography. 55th Thousand, revised to the present time. Crown 8vo, pp. 300, 3s.

—— The Intermediate Geography. Intended as an Intermediate Book between the Author's 'Outlines of Geography' and 'Elements of Geography.' Sixteenth Edition, revised. Crown 8vo, pp. 238, 2s.

—— Outlines of Modern Geography. 188th Thousand, revised to the present time. 18mo, pp. 118, 1s.

—— First Steps in Geography. 105th Thousand. 18mo, pp. 56. Sewed, 4d.; cloth, 6d.

—— Elements of Physiography and Physical Geography. With Express Reference to the Instructions issued by the Science and Art Department. 30th Thousand, revised. Crown 8vo, 1s. 6d.

—— Facts and Dates; or, the Leading Events in Sacred and Profane History, and the Principal Facts in the various Physical Sciences. For Schools and Private Reference. New Edition. Crown 8vo, 3s. 6d.

MACKAY. An Old Scots Brigade. Being the History of Mackay's Regiment, now incorporated with the Royal Scots. With an Appendix containing many Original Documents connected with the History of the Regiment. By JOHN MACKAY (late) OF HERRIESDALE. Crown 8vo, 5s.

MACKENZIE. Studies in Roman Law. With Comparative Views of the Laws of France, England, and Scotland. By LORD MACKENZIE, one of the Judges of the Court of Session in Scotland. Sixth Edition, Edited by JOHN KIRKPATRICK, Esq., M.A., LL.B., Advocate, Professor of History in the University of Edinburgh. 8vo, 12s.

M'KERLIE. Galloway: Ancient and Modern. An Account of the Historic Celtic District. By P. H. M'KERLIE, F.S.A. Scot., F.R.G.S., &c. Author of 'Lands and their Owners in Galloway.' Crown 8vo, 7s. 6d.

M'PHERSON. Summer Sundays in a Strathmore Parish. By J. GORDON M'PHERSON, Ph.D., F.R.S.E., Minister of Ruthven. Crown 8vo, 5s.

—— Golf and Golfers. Past and Present. With an Introduction by the Right Hon. A. J. BALFOUR, and a Portrait of the Author. Fcap. 8vo, 1s. 6d.

MAIN. Three Hundred English Sonnets. Chosen and Edited by DAVID M. MAIN. Fcap. 8vo, 6s.

MAIR. A Digest of Laws and Decisions, Ecclesiastical and Civil, relating to the Constitution, Practice, and Affairs of the Church of Scotland. With Notes and Forms of Procedure. By the Rev. WILLIAM MAIR, D.D., Minister of the Parish of Earlston. Crown 8vo. With Supplements, 8s.

MARMORNE. The Story is told by ADOLPHUS SEGRAVE, the youngest of three Brothers. Third Edition. Crown 8vo, 6s.

MARSHALL. French Home Life. By FREDERIC MARSHALL, Author of 'Claire Brandon.' Second Edition. 5s.

———— It Happened Yesterday. A Novel. Crown 8vo, 6s.

MARSHMAN. History of India. From the Earliest Period to the Close of the India Company's Government; with an Epitome of Subsequent Events. By JOHN CLARK MARSHMAN, C.S.I. Abridged from the Author's larger work. Second Edition, revised. Crown 8vo, with Map, 6s. 6d.

MARTIN. Goethe's Faust. Part I. Translated by Sir THEODORE MARTIN, K.C.B. Second Ed., crown 8vo, 6s. Ninth Ed., fcap. 8vo, 3s. 6d.

———— Goethe's Faust. Part II. Translated into English Verse. Second Edition, revised. Fcap. 8vo, 6s.

———— The Works of Horace. Translated into English Verse, with Life and Notes. 2 vols. New Edition, crown 8vo, 21s.

———— Poems and Ballads of Heinrich Heine. Done into Eng-lish Verse. Second Edition. Printed on *papier vergé*, crown 8vo, 8s.

———— The Song of the Bell, and other Translations from Schiller, Goethe, Uhland, and Others. Crown 8vo, 7s. 6d.

———— Catullus. With Life and Notes. Second Ed., post 8vo, 7s. 6d.

———— Aladdin : A Dramatic Poem. By ADAM OEHLENSCHLAE-GER. Fcap. 8vo, 5s.

———— Correggio : A Tragedy. By OEHLENSCHLAEGER. With Notes. Fcap. 8vo, 3s.

———— King Rene's Daughter : A Danish Lyrical Drama. By HENRIK HERTZ. Second Edition, fcap., 2s. 6d.

MARTIN. On some of Shakespeare's Female Characters. In a Series of Letters. By HELENA FAUCIT, LADY MARTIN. Dedicated by per-mission to Her Most Gracious Majesty the Queen. New Edition, enlarged. 8vo, with Portrait by Lane, 7s. 6d.

MATHESON. Can the Old Faith Live with the New? or the Problem of Evolution and Revelation. By the Rev. GEORGE MATHESON, D.D. Third Edition. Crown 8vo, 7s. 6d.

———— The Psalmist and the Scientist ; or, Modern Value of the Religious Sentiment. New and Cheaper Edition. Crown 8vo, 5s.

———— Spiritual Development of St Paul. 3d Edition. Cr. 8vo, 5s.

———— Sacred Songs. New and Cheaper Edition. Cr. 8vo, 2s. 6d.

MAURICE. The Balance of Military Power in Europe. An Examination of the War Resources of Great Britain and the Continental States. By Colonel MAURICE, R.A., Professor of Military Art and History at the Royal Staff College. Crown 8vo, with a Map. 6s.

MEREDYTH. The Brief for the Government, 1886-92. A Hand-book for Conservative and Unionist Writers, Speakers, &c. Second Edition. By W. H. MEREDYTH. Crown 8vo, 2s. 6d.

MICHEL. A Critical Inquiry into the Scottish Language. With the view of Illustrating the Rise and Progress of Civilisation in Scotland. By FRANCISQUE-MICHEL, F.S.A. Lond. and Scot., Correspondant de l'Institut de France, &c. 4to, printed on hand-made paper, and bound in Roxburghe, 66s.

MICHIE. The Larch : Being a Practical Treatise on its Culture and General Management. By CHRISTOPHER Y. MICHIE, Forester, Cullen House. Crown 8vo, with Illustrations. New and Cheaper Edition, enlarged, 5s.

———— The Practice of Forestry. Cr. 8vo, with Illustrations. 6s.

MIDDLETON. The Story of Alastair Bhan Comyn ; or, The Tragedy of Dunphail. A Tale of Tradition and Romance. By the Lady MIDDLETON. Square 8vo 10s. Cheaper Edition, 5s.

MILLER. Landscape Geology. A Plea for the Study of Geology by Landscape Painters. By HUGH MILLER, of H.M. Geological Survey. Cr. 8vo, 3s.

MILNE-HOME. Mamma's Black Nurse Stories. West Indian Folk-lore. By MARY PAMELA MILNE-HOME. With six full-page tinted Illus-trations. Small 4to, 5s.

MINTO.   A Manual of English Prose Literature, Biographical
and Critical : designed mainly to show Characteristics of Style. By W. MINTO,
M.A., Professor of Logic in the University of Aberdeen. Third Edition,
revised. Crown 8vo, 7s. 6d.
———— Characteristics of English Poets, from Chaucer to Shirley.
New Edition, revised. Crown 8vo, 7s. 6d.

MOIR.   Life of Mansie Wauch, Tailor in Dalkeith.   By D. M.
MOIR. With 8 Illustrations on Steel, by the late GEORGE CRUIKSHANK.
Crown 8vo, 3s. 6d. Another Edition, fcap. 8vo, 1s. 6d.

MOMERIE.   Defects of Modern Christianity, and other Sermons.
By ALFRED WILLIAMS MOMERIE, M.A., D.Sc., LL.D. 4th Edition. Cr. 8vo, 5s.
———— The Basis of Religion. Being an Examination of Natural
Religion. Third Edition. Crown 8vo, 2s. 6d.
———— The Origin of Evil, and other Sermons.   Seventh Edition,
enlarged. Crown 8vo, 5s.
———— Personality. The Beginning and End of Metaphysics, and
a Necessary Assumption in all Positive Philosophy. Fourth Ed. Cr. 8vo, 3s.
———— Agnosticism. Third Edition, Revised. Crown 8vo, 5s.
———— Preaching and Hearing; and other Sermons.   Third
Edition, Enlarged. Crown 8vo, 5s.
———— Belief in God. Third Edition. Crown 8vo, 3s.
———— Inspiration; and other Sermons. Second Ed. Cr. 8vo, 5s.
———— Church and Creed. Second Edition. Crown 8vo, 4s. 6d.

MONTAGUE.   Campaigning in South Africa.   Reminiscences of
an Officer in 1879. By Captain W. E. MONTAGUE, 94th Regiment, Author of
'Claude Meadowleigh,' &c. 8vo, 10s. 6d.

MONTALEMBERT.   Memoir of Count de Montalembert.   A
Chapter of Recent French History. By Mrs OLIPHANT, Author of the 'Life
of Edward Irving,' &c. 2 vols. crown 8vo, £1, 4s.

MORISON.   Sordello.   An Outline Analysis of Mr Browning's
Poem. By JEANIE MORISON, Author of 'The Purpose of the Ages,' 'Ane
Booke of Ballades,' &c. Crown 8vo, 3s.
———— Selections from Poems. Crown 8vo, 4s. 6d.
———— There as Here. Crown 8vo, 3s.
      ₊ A limited impression on handmade paper, bound in vellum, 7s. 6d.
———— "Of Fifine at the Fair," "Christmas Eve and Easter Day,"
and other of Mr Browning's Poems. Crown 8vo, 3s.

MOZLEY.   Essays from 'Blackwood.'   By the late ANNE MOZLEY,
Author of 'Essays on Social Subjects'; Editor of 'The Letters and Correspond-
ence of Cardinal Newman,' 'Letters of the Rev. J. B. Mozley,' &c. With a
Memoir by her Sister, FANNY MOZLEY. Post 8vo, 7s. 6d.

MUNRO.   On Valuation of Property.   By WILLIAM MUNRO, M.A.,
Her Majesty's Assessor of Railways and Canals for Scotland. Second Edition.
Revised and enlarged. 8vo, 3s. 6d.

MURDOCH.   Manual of the Law of Insolvency and Bankruptcy :
Comprehending a Summary of the Law of Insolvency, Notour Bankruptcy,
Composition - contracts, Trust-deeds, Cessios, and Sequestrations; and the
Winding-up of Joint-Stock Companies in Scotland; with Annotations on the
various Insolvency and Bankruptcy Statutes; and with Forms of Procedure
applicable to these Subjects. By JAMES MURDOCH, Member of the Faculty of
Procurators in Glasgow. Fifth Edition, Revised and Enlarged, 8vo, £1, 10s.

MY TRIVIAL LIFE AND MISFORTUNE : A Gossip with
no Plot in Particular. By A PLAIN WOMAN. Cheap Ed., crown 8vo, 3s. 6d.
            By the SAME AUTHOR.
      POOR NELLIE. New Edition. Crown 8vo, 6s.

NAPIER.   The Construction of the Wonderful Canon of Logar-
ithms. By JOHN NAPIER of Merchiston. Translated, with Notes, and a
Catalogue of Napier's Works, by WILLIAM RAE MACDONALD. Small 4to, 15s.
A few large-paper copies on Whatman paper, 30s.

NEAVES. Songs and Verses, Social and Scientific. By an Old
Contributor to 'Maga.' By the Hon. Lord NEAVES. Fifth Ed., fcap. 8vo, 4s.
—— The Greek Anthology. Being Vol. XX. of 'Ancient
Classics for English Readers.' Crown 8vo, 2s. 6d.

NICHOLSON. A Manual of Zoology, for the Use of Students.
With a General Introduction on the Principles of Zoology. By HENRY AL-
LEYNE NICHOLSON, M.D., D.Sc., F.L.S., F.G.S., Regius Professor of Natural
History in the University of Aberdeen. Seventh Edition, rewritten and
enlarged. Post 8vo, pp. 956, with 555 Engravings on Wood, 18s.
—— Text-Book of Zoology, for the Use of Schools. Fourth Edi-
tion, enlarged. Crown 8vo, with 188 Engravings on Wood, 7s. 6d.
—— Introductory Text-Book of Zoology, for the Use of Junior
Classes. Sixth Edition, revised and enlarged, with 166 Engravings, 3s.
—— Outlines of Natural History, for Beginners ; being Descrip-
tions of a Progressive Series of Zoological Types. Third Edition, with
Engravings, 1s. 6d.
—— A Manual of Palæontology, for the Use of Students.
With a General Introduction on the Principles of Palæontology. By Professor
H. ALLEYNE NICHOLSON and RICHARD LYDEKKER, B.A. Third Edition. Re-
written and greatly enlarged. 2 vols. 8vo, with Engravings, £3, 3s.
—— The Ancient Life-History of the Earth. An Outline of
the Principles and Leading Facts of Palæontological Science. Crown 8vo,
with 276 Engravings, 10s. 6d.
—— On the "Tabulate Corals" of the Palæozoic Period, with
Critical Descriptions of Illustrative Species. Illustrated with 15 Litho-
graph Plates and numerous Engravings. Super-royal 8vo, 21s.
—— Synopsis of the Classification of the Animal King-
dom. 8vo, with 106 Illustrations, 6s.
—— On the Structure and Affinities of the Genus Monticuli-
pora and its Sub-Genera, with Critical Descriptions of Illustrative Species.
Illustrated with numerous Engravings on wood and lithographed Plates.
Super-royal 8vo, 18s.

NICHOLSON. Communion with Heaven, and other Sermons.
By the late MAXWELL NICHOLSON, D.D., Minister of St Stephen's, Edinburgh.
Crown 8vo, 5s. 6d.
—— Rest in Jesus. Sixth Edition. Fcap. 8vo, 4s. 6d.

NICHOLSON. A Treatise on Money, and Essays on Present
Monetary Problems. By JOSEPH SHIELD NICHOLSON, M.A., D.Sc., Professor
of Commercial and Political Economy and Mercantile Law in the University
of Edinburgh. 8vo, 10s. 6d.
—— Thoth. A Romance. Third Edition. Crown 8vo, 4s. 6d.
—— A Dreamer of Dreams. A Modern Romance. Second
Edition. Crown 8vo, 6s.

NICOLSON AND MURE. A Handbook to the Local Govern-
ment (Scotland) Act, 1889. With Introduction, Explanatory Notes, and
Index. By J. BADENACH NICOLSON, Advocate, Counsel to the Scotch Education Department, and W. J. MURE, Advocate, Legal Secretary to the Lord
Advocate for Scotland. Ninth Reprint. 8vo, 5s.

OLIPHANT. Masollam : a Problem of the Period. A Novel.
By LAURENCE OLIPHANT. 3 vols. post 8vo, 25s. 6d
—— Scientific Religion ; or, Higher Possibilities of Life and
Practice through the Operation of Natural Forces. Second Edition. 8vo, 16s.
—— Altiora Peto. By LAURENCE OLIPHANT. Cheap Edition.
Crown 8vo, boards, 2s. 6d.; cloth, 3s. 6d. Illustrated Edition. Crown 8vo,
cloth, 6s.
—— Piccadilly. With Illustrations by Richard Doyle. New
Edition, 3s. 6d. Cheap Edition, boards. 2s. 6d.
—— Traits and Travesties ; Social and Political. Post 8vo, 10s. 6d.
—— Haifa : Life in Modern Palestine. 2d Edition. 8vo, 7s. 6d.
—— Episodes in a Life of Adventure ; or, Moss from a Rolling
Stone. Fifth Edition. Post 8vo, 6s.

OLIPHANT.   The Land of Gilead.   With Excursions in the
Lebanon.   With Illustrations and Maps.   Demy 8vo, 21s.
———— Memoir of the Life of Laurence Oliphant, and of Alice
Oliphant, his Wife.   By Mrs M. O. W. OLIPHANT.   Seventh Edition.   In 2 vols.
post 8vo, with Portraits.   21s.
POPULAR EDITION.   With a New Preface.   Post 8vo.   With Portraits, 7s. 6d.
OLIPHANT.   Katie Stewart.   By Mrs OLIPHANT.   2s. 6d.
———— Two Stories of the Seen and the Unseen.   The Open Door
—Old Lady Mary.   Paper Covers, 1s.
———— Sons and Daughters.   Crown 8vo, 3s. 6d.
OLIPHANT.   Notes of a Pilgrimage to Jerusalem and the Holy
Land.   By F. R. OLIPHANT.   Crown 8vo, 3s. 6d.
ON SURREY HILLS.   By "A SON OF THE MARSHES."   Second
Edition.   Crown 8vo, 6s.
BY THE SAME AUTHOR.
Annals of a Fishing Village.   Edited by J. A. OWEN.
Cheap Edition.   Crown 8vo, 6s.   With 7 full-page Illustrations, 7s. 6d.
Within an Hour of London Town.   Among Wild Birds and
their Haunts.   Crown 8vo, 6s.
OSSIAN.   The Poems of Ossian in the Original Gaelic.   With a
Literal Translation into English, and a Dissertation on the Authenticity of the
Poems.   By the Rev. ARCHIBALD CLERK.   2 vols. imperial 8vo, £1, 11s. 6d.
OSWALD.   By Fell and Fjord ; or, Scenes and Studies in Iceland.
By E. J. OSWALD.   Post 8vo, with Illustrations.   7s. 6d.
PAGE.   Introductory Text-Book of Geology.   By DAVID PAGE,
LL.D., Professor of Geology in the Durham University of Physical Science
Newcastle, and Professor LAPWORTH of Mason Science College, Birmingham.
With Engravings and Glossarial Index.   Twelfth Edition.   Revised and En-
larged.   3s. 6d.
———— Advanced Text-Book of Geology, Descriptive and Indus-
trial.   With Engravings, and Glossary of Scientific Terms.   Sixth Edition, re-
vised and enlarged, 7s. 6d.
———— Introductory Text-Book of Physical Geography.   With
Sketch-Maps and Illustrations.   Edited by Professor LAPWORTH, LL.D., F.G.S.,
&c., Mason Science College, Birmingham.   12th Edition.   2s. 6d.
———— Advanced Text-Book of Physical Geography.   Third
Edition, Revised and Enlarged by Prof. LAPWORTH.   With Engravings.   5s.
PATON.   Spindrift.   By Sir J. NOEL PATON.   Fcap., cloth, 5s.
———— Poems by a Painter.   Fcap., cloth, 5s.
PATON.   Body and Soul.   A Romance in Transcendental Path-
ology.   By FREDERICK NOEL PATON.   Third Edition.   Crown 8vo, 1s.
PATRICK.   The Apology of Origen in Reply to Celsus.   A
Chapter in the History of Apologetics.   By Rev. J. PATRICK, B.D.   In 1
vol. crown 8vo.   [In the press.
PATTERSON.   Essays in History and Art.   By R. HOGARTH
PATTERSON.   8vo, 12s.
———— The New Golden Age, and Influence of the Precious
Metals upon the World.   2 vols. 8vo, 31s. 6d.
PAUL.   History of the Royal Company of Archers, the Queen's
Body-Guard for Scotland.   By JAMES BALFOUR PAUL, Advocate of the Scottish
Bar.   Crown 4to, with Portraits and other Illustrations.   £2, 2s.
PEILE.   Lawn Tennis as a Game of Skill.   With latest revised
Laws as played by the Best Clubs.   By Captain S. C. F. PEILE, B.S.C.   Cheaper
Edition, fcap. cloth, 1s.
PETTIGREW.   The Handy Book of Bees, and their Profitable
Management.   By A. PETTIGREW.   Fifth Edition, Enlarged, with Engrav-
ings.   Crown 8vo, 3s. 6d.
PHILIP.   The Function of Labour in the Production of Wealth.
By ALEXANDER PHILIP, LL.B., Edinburgh.   Crown 8vo, 3s. 6d.

PHILOSOPHICAL CLASSICS FOR ENGLISH READERS. Edited by WILLIAM KNIGHT, LL.D., Professor of Moral Philosophy, University of St Andrews. In crown 8vo volumes, with portraits, price 3s. 6d.
[For list of Volumes published, see page 2.

POLLOK. The Course of Time : A Poem. By ROBERT POLLOK, A.M. Small fcap. 8vo, cloth gilt, 2s. 6d. Cottage Edition, 32mo, 8d. The Same, cloth, gilt edges, 1s. 6d. Another Edition, with Illustrations by Birket Foster and others, fcap., cloth, 3s. 6d., or with edges gilt, 4s.

PORT ROYAL LOGIC. Translated from the French ; with Introduction, Notes, and Appendix. By THOMAS SPENCER BAYNES, LL.D., Professor in the University of St Andrews. Tenth Edition, 12mo, 4s.

POTTS AND DARNELL. Aditus Faciliores : An easy Latin Construing Book, with Complete Vocabulary. By the late A. W. POTTS, M.A., LL.D., and the Rev. C. DARNELL, M.A., Head-Master of Cargilfield Preparatory School, Edinburgh. Tenth Edition, fcap. 8vo, 3s. 6d.

——— Aditus Faciliores Graeci. An easy Greek Construing Book, with Complete Vocabulary. Fourth Edition, fcap. 8vo, 3s.

POTTS. School Sermons. By the late ALEXANDER WM. POTTS, LL.D., First Head-Master of Fettes College. With a Memoir and Portrait. Crown 8vo, 7s. 6d.

PRINGLE. The Live-Stock of the Farm. By ROBERT O. PRINGLE. Third Edition. Revised and Edited by JAMES MACDONALD. Cr. 8vo, 7s. 6d.

PUBLIC GENERAL STATUTES AFFECTING SCOTLAND from 1707 to 1847, with Chronological Table and Index. 3 vols. large 8vo, £3, 3s.

PUBLIC GENERAL STATUTES AFFECTING SCOTLAND, COLLECTION OF. Published Annually with General Index.

RADICAL CURE FOR IRELAND, The. A Letter to the People of England and Scotland concerning a new Plantation. With 2 Maps. 8vo, 7s. 6d.

RAE. The Syrian Church in India. By GEORGE MILNE RAE, M.A., Fellow of the University of Madras ; late Professor in the Madras Christian College. With 6 full-page Illustrations. Post 8vo, 10s. 6d.

RAMSAY. Scotland and Scotsmen in the Eighteenth Century. Edited from the MSS. of JOHN RAMSAY, Esq. of Ochtertyre, by ALEXANDER ALLARDYCE, Author of 'Memoir of Admiral Lord Keith, K.B.,' &c. 2 vols. 8vo, 31s. 6d.

RANKIN. A Handbook of the Church of Scotland. By JAMES RANKIN, D.D., Minister of Muthill ; Author of 'Character Studies in the Old Testament,' &c. An entirely New and much Enlarged Edition. Crown 8vo, with 2 Maps, 7s. 6d.

——— The Creed in Scotland. An Exposition of the Apostles' Creed. With Extracts from Archbishop Hamilton's Catechism of 1552, John Calvin's Catechism of 1556, and a Catena of Ancient Latin and other Hymns. Post 8vo, 7s. 6d.

——— First Communion Lessons. 23d Edition. Paper Cover, 2d.

RECORDS OF THE TERCENTENARY FESTIVAL OF THE UNIVERSITY OF EDINBURGH. Celebrated in April 1884. Published under the Sanction of the Senatus Academicus. Large 4to, £2, 12s. 6d.

ROBERTSON. The Early Religion of Israel. As set forth by Biblical Writers and Modern Critical Historians. Being the Baird Lecture for 1888-89. By JAMES ROBERTSON, D.D., Professor of Oriental Languages in the University of Glasgow. Crown 8vo, 10s. 6d.

ROBERTSON. Orellana, and other Poems. By J. LOGIE ROBERTSON, M.A. Fcap. 8vo. Printed on hand-made paper. 6s.

ROBERTSON. Our Holiday Among the Hills. By JAMES and JANET LOGIE ROBERTSON. Fcap. 8vo, 3s. 6d.

ROSCOE. Rambles with a Fishing-rod. By E. S. ROSCOE. Crown 8vo, 4s. 6d.

ROSS. Old Scottish Regimental Colours. By ANDREW ROSS, S.S.C., Hon. Secretary Old Scottish Regimental Colours Committee. Dedicated by Special Permission to Her Majesty the Queen. Folio. £2, 12s. 6d.

RUSSELL. The Haigs of Bemersyde. A Family History. By JOHN RUSSELL. Large 8vo, with Illustrations. 21s.

RUSSELL. Fragments from Many Tables. Being the Recollections of some Wise and Witty Men and Women. By GEO. RUSSELL. Cr. 8vo, 4s. 6d.

RUTLAND. Notes of an Irish Tour in 1846. By the DUKE OF RUTLAND, G.C.B. (Lord JOHN MANNERS). New Edition. Crown 8vo, 2s. 6d.

———— Correspondence between the Right Honble. William Pitt and Charles Duke of Rutland, Lord Lieutenant of Ireland, 1781-1787. With Introductory Note by John Duke of Rutland. 8vo, 7s. 6d.

RUTLAND. Gems of German Poetry. Translated by the DUCHESS OF RUTLAND (Lady JOHN MANNERS).   [New Edition in preparation.

———— Impressions of Bad-Homburg. Comprising a Short Account of the Women's Associations of Germany under the Red Cross. Crown 8vo, 1s. 6d.

———— Some Personal Recollections of the Later Years of the Earl of Beaconsfield, K.G. Sixth Edition, 6d.

———— Employment of Women in the Public Service. 6d.

———— Some of the Advantages of Easily Accessible Reading and Recreation Rooms, and Free Libraries. With Remarks on Starting and Maintaining Them. Second Edition, crown 8vo, 1s.

———— A Sequel to Rich Men's Dwellings, and other Occasional Papers. Crown 8vo, 2s. 6d.

———— Encouraging Experiences of Reading and Recreation Rooms, Aims of Guilds, Nottingham Social Guild, Existing Institutions, &c., &c. Crown 8vo, 1s.

SCHILLER. Wallenstein. A Dramatic Poem. By FREDERICK VON SCHILLER. Translated by C. G. A. LOCKHART. Fcap. 8vo, 7s. 6d.

SCOTCH LOCH FISHING. By "Black Palmer." Crown 8vo, interleaved with blank pages, 4s.

SCOUGAL. Prisons and their Inmates; or, Scenes from a Silent World. By FRANCIS SCOUGAL. Crown 8vo, boards, 2s.

SELLAR. Manual of the Education Acts for Scotland. By the late ALEXANDER CRAIG SELLAR, M.P. Eighth Edition. Revised and in great part rewritten by J. EDWARD GRAHAM, B.A. Oxon., Advocate. With Rules for the conduct of Elections, with Notes and Cases. With a Supplement, being the Acts of 1889 in so far as affecting the Education Acts. 8vo, 12s. 6d.
[SUPPLEMENT TO SELLAR'S MANUAL OF THE EDUCATION ACTS. 8vo, 2s.]

SETH. Scottish Philosophy. A Comparison of the Scottish and German Answers to Hume. Balfour Philosophical Lectures, University of Edinburgh. By ANDREW SETH, M.A., Professor of Logic and Metaphysics in Edinburgh University. Second Edition. Crown 8vo, 5s.

———— Hegelianism and Personality. Balfour Philosophical Lectures. Second Series. Crown 8vo, 5s.

SETH. Freedom as Ethical Postulate. By JAMES SETH, M.A., George Munro Professor of Philosophy, Dalhousie College, Halifax, Canada. 8vo, 1s.

SHADWELL. The Life of Colin Campbell, Lord Clyde. Illustrated by Extracts from his Diary and Correspondence. By Lieutenant-General SHADWELL, C.B. 2 vols. 8vo. With Portrait, Maps, and Plans. 36s.

SHAND. Half a Century; or, Changes in Men and Manners. By ALEX. INNES SHAND, Author of 'Against Time,' &c. Second Edition, 8vo, 12s. 6d.

———— Letters from the West of Ireland. Reprinted from the 'Times.' Crown 8vo, 5s.

———— Kilcarra. A Novel. 3 vols. crown 8vo, 25s. 6d.

SHARPE. Letters from and to Charles Kirkpatrick Sharpe. Edited by ALEXANDER ALLARDYCE, Author of 'Memoir of Admiral Lord Keith, K.B.,' &c. With a Memoir by the Rev. W. K. R. BEDFORD. In two vols. 8vo. Illustrated with Etchings and other Engravings. £2, 12s. 6d.

SIM. Margaret Sim's Cookery. With an Introduction by L. B. WALFORD, Author of 'Mr Smith : A Part of His Life,' &c. Crown 8vo, 5s.

SKELTON. Maitland of Lethington ; and the Scotland of Mary Stuart. A History. By JOHN SKELTON, C.B., LL.D., Author of 'The Essays of Shirley.' Demy 8vo. 2 vols., 28s.

———— The Handbook of Public Health. A Complete Edition of the Public Health and other Sanitary Acts relating to Scotland. Annotated, and with the Rules, Instructions, and Decisions of the Board of Supervision brought up to date with relative forms. 8vo, with Supplement, 8s. 6d.

———— Supplement to Skelton's Handbook. The Administration of the Public Health Act in Counties. 8vo, cloth, 1s. 6d.

———— The Local Government (Scotland) Act in Relation to Public Health. A Handy Guide for County and District Councillors, Medical Officers, Sanitary Inspectors, and Members of Parochial Boards. Second Edition. With a new Preface on appointment of Sanitary Officers. Crown 8vo, 2s.

SMITH. For God and Humanity. A Romance of Mount Carmel. By HASKETT SMITH, Author of 'The Divine Epiphany,' &c. 3 vols. post 8vo, 25s. 6d.

SMITH. Thorndale ; or, The Conflict of Opinions. By WILLIAM SMITH, Author of 'A Discourse on Ethics,' &c. New Edition. Cr. 8vo, 10s. 6d.

———— Gravenhurst ; or, Thoughts on Good and Evil. Second Edition, with Memoir of the Author. Crown 8vo, 8s.

———— The Story of William and Lucy Smith. Edited by GEORGE MERRIAM. Large post 8vo, 12s. 6d.

SMITH. Memoir of the Families of M'Combie and Thoms, originally M'Intosh and M'Thomas. Compiled from History and Tradition. By WILLIAM M'COMBIE SMITH. With Illustrations. 8vo, 7s. 6d.

SMITH. Greek Testament Lessons for Colleges, Schools, and Private Students, consisting chiefly of the Sermon on the Mount and the Parables of our Lord. With Notes and Essays. By the Rev. J. HUNTER SMITH, M.A., King Edward's School, Birmingham. Crown 8vo, 6s.

SMITH. Writings by the Way. By JOHN CAMPBELL SMITH, M.A., Sheriff-Substitute. Crown 8vo, 9s.

SMITH. The Secretary for Scotland. Being a Statement of the Powers and Duties of the new Scottish Office. With a Short Historical Introduction and numerous references to important Administrative Documents. By W. C. SMITH, LL.B., Advocate. 8vo, 6s.

SORLEY. The Ethics of Naturalism. Being the Shaw Fellowship Lectures, 1884. By W. R. SORLEY, M.A., Fellow of Trinity College, Cambridge, Professor of Logic and Philosophy in University College of South Wales. Crown 8vo, 6s.

SPEEDY. Sport in the Highlands and Lowlands of Scotland with Rod and Gun. By TOM SPEEDY. Second Edition, Revised and Enlarged. With Illustrations by Lieut.-Gen. Hope Crealocke, C.B., C.M.G., and others. 8vo, 15s.

SPROTT. The Worship and Offices of the Church of Scotland. By GEORGE W. SPROTT, D.D., Minister of North Berwick. Crown 8vo, 6s.

STAFFORD. How I Spent my Twentieth Year. Being a Record of a Tour Round the World, 1886-87. By the MARCHIONESS OF STAFFORD. With Illustrations. Third Edition, crown 8vo, 8s. 6d.

STARFORTH. Villa Residences and Farm Architecture : A Series of Designs. By JOHN STARFORTH, Architect. 102 Engravings. Second Edition, medium 4to, £2, 17s. 6d.

STATISTICAL ACCOUNT OF SCOTLAND. Complete, with Index, 15 vols. 8vo, £16, 16s.
Each County sold separately, with Title, Index, and Map, neatly bound in cloth.

STEPHENS' BOOK OF THE FARM. Illustrated with numerous Portraits of Animals and Engravings of Implements, and Plans of Farm Buildings. Fourth Edition. Revised, and in great part rewritten by JAMES MACDONALD, of the 'Farming World,' &c. Complete in Six Divisional Volumes, bound in cloth, each 10s. 6d., or handsomely bound, in 3 volumes, with leather back and gilt top, £3, 3s.

———— The Book of Farm Implements and Machines. By J. SLIGHT and R. SCOTT BURN, Engineers. Edited by HENRY STEPHENS. Large 8vo, £2. 2s.

STEVENSON. British Fungi. (Hymenomycetes.) By Rev. JOHN STEVENSON, Author of 'Mycologia Scotia,' Hon. Sec. Cryptogamic Society of Scotland. Vols. I. and II., post 8vo. with Illustrations, price 12s. 6d. each.

STEWART. Advice to Purchasers of Horses. By JOHN STEWART, V.S. New Edition. 2s. 6d.

—— Stable Economy. A Treatise on the Management of Horses in relation to Stabling, Grooming, Feeding, Watering, and Working. Seventh Edition, fcap. 8vo, 6s. 6d.

STEWART. A Hebrew Grammar, with the Pronunciation, Syllabic Division and Tone of the Words, and Quantity of the Vowels. By Rev. DUNCAN STEWART, D.D. Fourth Edition. 8vo, 3s. 6d.

STEWART. Boethius : An Essay. By HUGH FRASER STEWART, M.A., Trinity College, Cambridge. Crown 8vo, 7s. 6d.

STODDART. Angling Songs. By THOMAS TOD STODDART. New Edition, with a Memoir by ANNA M. STODDART. Crown 8vo, 7s. 6d.

STORMONTH. Etymological and Pronouncing Dictionary of the English Language. Including a very Copious Selection of Scientific Terms. For Use in Schools and Colleges, and as a Book of General Reference. By the Rev. JAMES STORMONTH. The Pronunciation carefully Revised by the Rev. P. H. PHELP, M.A. Cantab. Tenth Edition, Revised throughout. Crown 8vo, pp. 800. 7s. 6d.

—— Dictionary of the English Language, Pronouncing, Etymological, and Explanatory. Revised by the Rev. P. H. PHELP. Library Edition. Imperial 8vo, handsomely bound in half morocco, 31s. 6d.

—— The School Etymological Dictionary and Word-Book. Fourth Edition. Fcap. 8vo, pp. 254. 2s.

STORY. Nero ; A Historical Play. By W. W. STORY, Author of 'Roba di Roma.' Fcap. 8vo, 6s.

—— Vallombrosa. Post 8vo, 5s.

—— Poems. 2 vols. fcap., 7s. 6d.

—— Fiammetta. A Summer Idyl. Crown 8vo, 7s. 6d.

—— Conversations in a Studio. 2 vols. crown 8vo, 12s. 6d.

—— Excursions in Art and Letters. Crown 8vo, 7s. 6d.

STRICKLAND. Life of Agnes Strickland. By her SISTER. Post 8vo, with Portrait engraved on Steel, 12s. 6d.

STURGIS. John - a - Dreams. A Tale. By JULIAN STURGIS. New Edition, crown 8vo, 3s. 6d.

—— Little Comedies, Old and New. Crown 8vo, 7s. 6d.

SUTHERLAND. Handbook of Hardy Herbaceous and Alpine Flowers, for general Garden Decoration. Containing Descriptions of upwards of 1000 Species of Ornamental Hardy Perennial and Alpine Plants; along with Concise and Plain Instructions for their Propagation and Culture. By WILLIAM SUTHERLAND, Landscape Gardener; formerly Manager of the Herbaceous Department at Kew. Crown 8vo, 7s. 6d.

TAYLOR. The Story of My Life. By the late Colonel MEADOWS TAYLOR, Author of 'The Confessions of a Thug,' &c. &c. Edited by his Daughter. New and cheaper Edition, being the Fourth. Crown 8vo, 6s.

TELLET. Pastor and Prelate. A Story of Clerical Life. By ROY TELLET, Author of 'The Outcasts,' &c. 3 vols. crown 8vo, 25s. 6d.

THOLUCK. Hours of Christian Devotion. Translated from the German of A. Tholuck, D.D., Professor of Theology in the University of Halle. By the Rev. ROBERT MENZIES, D.D. With a Preface written for this Translation by the Author. Second Edition, crown 8vo, 7s. 6d.

THOMSON. Handy Book of the Flower-Garden : being Practical Directions for the Propagation, Culture, and Arrangement of Plants in Flower-Gardens all the year round. With Engraved Plans. By DAVID THOMSON, Gardener to his Grace the Duke of Buccleuch, K.T., at Drumlanrig Fourth and Cheaper Edition, crown 8vo, 5s.

—— The Handy Book of Fruit-Culture under Glass: being a series of Elaborate Practical Treatises on the Cultivation and Forcing of Pines, Vines, Peaches, Figs, Melons, Strawberries, and Cucumbers. With Engravings of Hothouses, &c. Second Ed. Cr. 8vo, 7s. 6d.

THOMSON. A Practical Treatise on the Cultivation of the Grape Vine. By WILLIAM THOMSON, Tweed Vineyards. Tenth Edition, 8vo, 5s.

THOMSON. Cookery for the Sick and Convalescent. With Directions for the Preparation of Poultices, Fomentations, &c. By BARBARA THOMSON. Fcap. 8vo, 1s. 6d.

THORNTON. Opposites. A Series of Essays on the Unpopular Sides of Popular Questions. By LEWIS THORNTON. 8vo, 12s. 6d.

TOM CRINGLE'S LOG. A New Edition, with Illustrations. Crown 8vo, cloth gilt, 5s. Cheap Edition, 2s.

TRANSACTIONS OF THE HIGHLAND AND AGRICUL-TURAL SOCIETY OF SCOTLAND. Published annually, price 5s.

TULLOCH. Rational Theology and Christian Philosophy in England in the Seventeenth Century. By JOHN TULLOCH, D.D., Principal of St Mary's College in the University of St Andrews; and one of her Majesty's Chaplains in Ordinary in Scotland. Second Edition. 2 vols. 8vo, 16s.

—— Modern Theories in Philosophy and Religion. 8vo, 15s.

—— Luther, and other Leaders of the Reformation. Third Edition, enlarged. Crown 8vo, 3s. 6d.

—— Memoir of Principal Tulloch, D.D., LL.D. By Mrs OLIPHANT, Author of 'Life of Edward Irving.' Third and Cheaper Edition. 8vo, with Portrait. 7s. 6d.

TWEEDIE. The Arabian Horse : his Country and People. With Portraits of Typical or Famous Arabians, and numerous other Illustrations : also a Map of the Country of the Arabian Horse, and a descriptive Glossary of Arabic words and proper names. By Colonel W. TWEEDIE, C.S I., Bengal Staff Corps, H.B.M.'s Consul-General, Baghdad. [In the press.

VEITCH. Institutes of Logic. By JOHN VEITCH, LL.D., Professor of Logic and Rhetoric in the University of Glasgow. Post 8vo, 12s. 6d.

—— The Feeling for Nature in Scottish Poetry. From the Earliest Times to the Present Day. 2 vols. fcap. 8vo, in roxburghe binding. 15s.

—— Merlin and Other Poems. Fcap. 8vo. 4s. 6d.

—— Knowing and Being. Essays in Philosophy. First Series. Crown 8vo, 5s.

VIRGIL. The Æneid of Virgil. Translated in English Blank Verse by G. K. RICKARDS, M.A., and Lord RAVENSWORTH. 2 vols. fcap. 8vo, 10s.

WALFORD. Four Biographies from 'Blackwood': Jane Taylor, Hannah More, Elizabeth Fry, Mary Somerville. By L. B. WALFORD. Crown 8vo, 5s.

WARREN'S (SAMUEL) WORKS :—
Diary of a Late Physician. Cloth, 2s. 6d.; boards, 2s.
Ten Thousand A-Year. Cloth, 3s. 6d.; boards, 2s. 6d.
Now and Then. The Lily and the Bee. Intellectual and Moral Development of the Present Age. 4s. 6d.
Essays : Critical, Imaginative, and Juridical. 5s.

WARREN. The Five Books of the Psalms. With Marginal Notes. By Rev. SAMUEL L. WARREN, Rector of Esher, Surrey; late Fellow, Dean, and Divinity Lecturer, Wadham College, Oxford. Crown 8vo, 5s.

WEBSTER. The Angler and the Loop-Rod. By DAVID WEBSTER. Crown 8vo, with Illustrations, 7s. 6d.

WELLINGTON. Wellington Prize Essays on "the System of Field Manœuvres best adapted for enabling our Troops to meet a Continental Army.' Edited by General Sir EDWARD BRUCE HAMLEY, K.C.B., K.C.M.G. 8vo, 12s. 6d.

WENLEY. Socrates and Christ : A Study in the Philosophy of Religion. By R. M. WENLEY, M.A., Lecturer on Mental and Moral Philosophy in Queen Margaret College, Glasgow; Examiner in Philosophy in the University of Glasgow. Crown 8vo, 6s.

WERNER. A Visit to Stanley's Rear-Guard at Major Bartte-
lot's Camp on the Aruhwimi. With an Account of River-Life on the Congo.
By J. R. WERNER, F.R.G.S., Engineer, late in the Service of the Etat Inde-
pendant du Congo. With Maps, Portraits, and other Illustrations. 8vo. 16s.

WESTMINSTER ASSEMBLY. Minutes of the Westminster As-
sembly, while engaged in preparing their Directory for Church Government,
Confession of Faith, and Catechisms (November 1644 to March 1649). Edited
by the Rev. Professor ALEX. T. MITCHELL, of St Andrews, and the Rev. JOHN
STRUTHERS, LL.D. With a Historical and Critical Introduction by Professor
Mitchell. 8vo, 15s.

WHITE. The Eighteen Christian Centuries. By the Rev. JAMES
WHITE. Seventh Edition, post 8vo, with Index, 6s.

———— History of France, from the Earliest Times. Sixth Thousand, post 8vo, with Index, 6s.

WHITE. Archæological Sketches in Scotland—Kintyre and Knap-
dale. By Colonel T. P. WHITE, R.E., of the Ordnance Survey. With numerous
Illustrations. 2 vols. folio, £4, 4s. Vol. I., Kintyre, sold separately, £2, 2s.

———— The Ordnance Survey of the United Kingdom. A Popular
Account. Crown 8vo, 5s.

WICKS. Golden Lives. The Story of a Woman's Courage. By
FREDERICK WICKS. Cheap Edition, with 120 Illustrations. Illustrated
Boards. 8vo, 2s. 6d.

WILLIAMSON. The Horticultural Exhibitors' Handbook. A
Treatise on Cultivating, Exhibiting, and Judging Plants, Flowers, Fruits, and
Vegetables. By W. WILLIAMSON, Gardener. Revised by MALCOLM DUNN, Gar-
dener to His Grace the Duke of Buccleuch and Queensberry, Dalkeith Park.
Cr. 8vo, 3s. 6d.

WILLIAMSON. Poems of Nature and Life. By DAVID R.
WILLIAMSON, Minister of Kirkmaiden. Fcap. 8vo, 3s.

WILLS AND GREENE. Drawing-room Dramas for Children. By
W. G. WILLS and the Hon. Mrs GREENE. Crown 8vo, 6s.

WILSON. Works of Professor Wilson. Edited by his Son-in-Law,
Professor FERRIER. 12 vols. crown 8vo, £2, 8s.

———— Christopher in his Sporting-Jacket. 2 vols., 8s.

———— Isle of Palms, City of the Plague, and other Poems. 4s.

———— Lights and Shadows of Scottish Life, and other Tales. 4s.

———— Essays, Critical and Imaginative. 4 vols., 16s.

———— The Noctes Ambrosianæ. 4 vols., 16s.   [8vo, 4s.

———— Homer and his Translators, and the Greek Drama. Crown

WINGATE. Lily Neil. A Poem. By DAVID WINGATE. Crown
8vo, 4s. 6d.

WORDSWORTH. The Historical Plays of Shakspeare. With
Introductions and Notes. By CHARLES WORDSWORTH, D.C.L., Bishop of S.
Andrews. 3 vols. post 8vo, cloth, each price 7s. 6d., or handsomely bound in
half-calf, each price 9s. 9d.

WORSLEY. Poems and Translations. By PHILIP STANHOPE
WORSLEY, M.A. Edited by EDWARD WORSLEY. 2d Ed., enlarged. Fcap. 8vo, 6s.

YATE. England and Russia Face to Face in Asia. A Record of
Travel with the Afghan Boundary Commission. By Captain A. C. YATE,
Bombay Staff Corps. 8vo, with Maps and Illustrations, 21s.

YATE. Northern Afghanistan; or, Letters from the Afghan
Boundary Commission. By Major C. E. YATE, C.S.I., C.M.G. Bombay Staff
Corps, F.R.G.S. 8vo, with Maps. 18s.

YOUNG. A Story of Active Service in Foreign Lands. Compiled
from letters sent home from South Africa, India, and China, 1856-1882. By
Surgeon-General A. GRAHAM YOUNG, Author of 'Crimean Cracks.' Crown
8vo, Illustrated, 7s. 6d.

YULE. Fortification: for the Use of Officers in the Army, and
Readers of Military History. By Col. YULE, Bengal Engineers. 8vo, with
numerous Illustrations, 10s. 6d.

4/92.

CPSIA information can be obtained
at www.ICGtesting.com
Printed in the USA
LVHW052210120723
752162LV00006B/422

9 781014 959102